# Exterior Calculus:
# Theory and Cases

## Authored by

## Carlos Polanco

*Faculty of Sciences*
*Universidad Nacional Autónoma de México*
*México*

# Exterior Calculus: Theory and Cases

Author: Carlos Polanco

ISBN (Online): 978-981-4998-78-9

ISBN (Print): 978-981-4998-79-6

ISBN (Paperback): 978-981-4998-80-2

need for a court order if at any point you breach any terms of this License Agreement. In no event will any delay or failure by Bentham Science Publishers in enforcing your compliance with this License Agreement constitute a waiver of any of its rights.

3. You acknowledge that you have read this License Agreement, and agree to be bound by its terms and conditions. To the extent that any other terms and conditions presented on any website of Bentham Science Publishers conflict with, or are inconsistent with, the terms and conditions set out in this License Agreement, you acknowledge that the terms and conditions set out in this License Agreement shall prevail.

**Bentham Science Publishers Pte. Ltd.**
80 Robinson Road #02-00
Singapore 068898
Singapore
Email: subscriptions@benthamscience.net

**BENTHAM SCIENCE**

# CONTENTS

# FOREWORD

I congratulate Carlos Polanco for his experienced and insightful book Exterior Calculus – Theory and Cases. This work covers profoundly advanced Calculus for a readership that has acquired the necessary mathematical comprehension to direct to geometric algebra. It will guide higher level students as well as their teachers straightforwardly through this topic from Heaviside-Gibbs algebra over Grassmann algebra to differentiation, integration and fundamental theorems of Calculus. Despite the complexity of the subject, this book is written in a highly didactic style, which is reflecting the expertise and the long-term teaching experience of the author at the Universidad Nacional Autónoma de México.

The presentation of many examples and case studies as well the solution guide to the chapter exercises at the end of this book will help the readers to deepen and to inspect their acquired knowledge and to relate the theory to practice. I wish that Carlos Polanco's book will become part of many bookshelves and highly recommend it as a solid and distinctive textbook for advanced courses in Calculus.

**Thomas Buhse**
Universidad Autónoma del Estado de Morelos
Cuernavaca Morelos, Mexico.

# PREFACE

This Exterior Calculus ebook has been designed for third-year students of Sciences, as it contains the fundamentals related to Geometric algebra or Grassmann algebra oriented to Calculus. Without any doubt, this algebra has important implications in Science and Engineering. Here, the reader will find a clear presentation of the Geometric algebra on a plane and in space, as well as the extension of all its operators in $\mathbb{R}^n$. In order to make the comprehension of this important algebra easier, some examples and completely solved exercises are included.

The ebook thoroughly examines the elements of Geometric algebra $G$ over the Real **field** and these operators: inner product, outer product, and geometric product, their components, and their geometric representation, as well as their properties and the rigid transformations on the plane and in space. It also reviews the **differentiation** and the **integration** over Geometric algebra, including the line integral and surface integral. The Green, Stokes and Gauss theorems are also studied in detail and the Theorem of Fundamental Calculus is generalized.

The author hopes the reader interested in the study of the fundamentals of Exterior calculus, finds useful the material presented here and that the students that start studying this field find this information motivating. The author would like to acknowledge the Faculty of Sciences at Universidad Nacional Autónoma de México for support.

## CONFLICT OF INTEREST

The author declares no conflict of interest regarding the contents of each of the chapters of this ebook.

## CONSENT FOR PUBLICATION

Not applicable.

**Carlos Polanco**

Faculty of Sciences
Universidad Nacional Autónoma de México
México

&

Department of Electromechanical Instrumentation
Instituto Nacional de Cardiología Ignacio Chávez
México

# ACKNOWLEDGEMENTS

I would like to thank all those whose recommendations made possible the publication of this ebook.

# DEDICATION

The beauty of mathematics only
shows itself to more patient
followers.

*Maryam Mirzakhani*

# List of Credits

# List of Symbols

| Symbol | Description | Page |
|---|---|---|
| $<a>$ | Blades on $\mathbb{G}_2$ | 28 |
| $\lVert a \rVert$ | Norm on $\mathbb{G}_2$ | 28 |
| $v_{\parallel}$ | Parallel component of vector $v$ on $\mathbb{G}_2$ | 29 |
| $v_{\perp}$ | Perpendicular component of vector $v$ on $\mathbb{G}_2$ | 29 |
| $I = \sigma_1\sigma_2$ | Pseudo-vector on $\mathbb{G}_2$ | 29 |
| $Ia$ | Clockwise rotation on $\mathbb{G}_2$ | 29 |
| $aI$ | Counter-clockwise rotation on $\mathbb{G}_2$ | 29 |
| $(\mathbf{x}-x_0)\wedge v = 0$ | Equation of a line on $\mathbb{G}_2$ | 33 |
| $\mathbb{G}_3$ | Geometric algebra on $\mathbb{R}^3$ | 37 |
| $\mathbb{G}_3$ | Grassmann algebra on $\mathbb{R}^3$ | 37 |
| $a\wedge b$ | Outer product on $\mathbb{G}_3$ | 38 |
| $a\cdot b$ | Inner product on $\mathbb{G}_3$ | 38 |
| $ab$ | Geometric product on $\mathbb{G}_3$ | 39 |
| $\sigma_1\sigma_2\sigma_3$ | Trivector | 40 |
| $a(b+c)$ | Distributivity of geometric product $\mathbb{G}_3$ | 43 |
| $a\wedge(b+c)$ | Distributivity of outer product $\mathbb{G}_3$ | 43 |
| $a^{-1}$ | Multiplicative inverse on $\mathbb{G}_3$ | 44 |
| $a(bc) = (ab)c$ | Associativity on $\mathbb{G}_3$ | 44 |
| $a^{\dagger}$ | Reversion on $\mathbb{G}_3$ | 45 |
| $Ia_r$ | Dual on $\mathbb{G}_3$ | 45 |
| $<a>$ | Blades on $\mathbb{G}_3$ | 46 |
| $\lVert a \rVert$ | Norm on $\mathbb{G}_3$ | 47 |
| $v_{\parallel}$ | Parallel component of vector $v$ on $\mathbb{G}_3$ | 47 |
| $v_{\perp}$ | Perpendicular component of vector $v$ on $\mathbb{G}_3$ | 47 |
| $I = \sigma_1\sigma_2\sigma_3$ | Pseudo-vector on $\mathbb{G}_3$ | 48 |
| $Ia$ | Clockwise rotation on $\mathbb{G}_3$ | 48 |
| $aI$ | Counter-clockwise rotation on $\mathbb{G}_3$ | 48 |
| $(\mathbf{x}-x_0)\wedge v = 0$ | Equation of a line on $\mathbb{G}_3$ | 50 |
| $a\wedge b$ | Outer product on $\mathbb{G}_n$ | 56 |
| $a\cdot b$ | Inner product on $\mathbb{G}_n$ | 57 |
| $ab$ | Geometric product on $\mathbb{G}_n$ | 57 |
| $\sigma_1\sigma_2\cdots\sigma_n$ | Multivector | 58 |
| $a(b+c)$ | Distributivity of geometric product $\mathbb{G}_n$ | 58 |
| $a\wedge(b+c)$ | Distributivity of outer product $\mathbb{G}_n$ | 58 |
| $a^{-1}$ | Multiplicative inverse on $\mathbb{G}_n$ | 59 |
| $a(bc) = (ab)c$ | Associativity on $\mathbb{G}_n$ | 59 |
| $a^{\dagger}$ | Reversion on $\mathbb{G}_n$ | 60 |
| $Ia_r$ | Dual on $\mathbb{G}_n$ | 60 |
| $<a>$ | Blades on $\mathbb{G}_n$ | 61 |
| $\lVert a \rVert$ | Norm on $\mathbb{G}_n$ | 61 |
| $v_{\parallel}$ | Parallel component of vector $v$ on $\mathbb{G}_n$ | 62 |
| $v_{\perp}$ | Perpendicular component of vector $v$ on $\mathbb{G}_n$ | 62 |
| $I = \sigma_1\sigma_2\cdots\sigma_n$ | Pseudo-vector on $\mathbb{G}_n$ | 62 |
| $Ia$ | Clockwise rotation on $\mathbb{G}_n$ | 62 |
| $aI$ | Counter-clockwise rotation on $\mathbb{G}_n$ | 62 |
| $(\mathbf{x}-x_0)\wedge v = 0$ | Equation of a line on $\mathbb{G}_n$ | 63 |
| $(\mathbf{x}-x_0)\wedge v = 0$ | Equation of a multivector on $\mathbb{G}_n$ | 64 |
| $dw$ | Outer derivative | 67 |
| $w_0$ | 0-form | 68 |

# Part I

# Heaviside-Gibbs Algebra

The operators of the 'bf Heaviside-Gibbs algebra' have a major role in Vector Calculus. The next chapter focuses on the definition of the main operators, showing its usefulness in solving problems in 2-dimensional and 3-dimensional space, and it also discusses the robustness and limitations of this algebra in n-dimensional space.

# CHAPTER 1

# Vector Algebra on $\mathbb{R}^2$ and $\mathbb{R}^3$

**Carlos Polanco**

Faculty of Sciences, Universidad Nacional Autónoma de México, México

**Abstract** In this chapter, we introduce the main operators of Heaviside-Gibbs algebra: addition, subtraction, norm of vectors, as well as inner and cross product. From the point of view of Vector Calculus, we introduce the line and surface integrals, and the Green's, Stokes', and Gauss' Theorems. The last section discusses the extension of this algebra in n-dimensional space. The examples are in plane and space.

**Keywords**: cross product: $v \times w$, divergence of vector function, Gauss' Theorem, Green's Theorem, inner product: $v \cdot w$, limitations, line integral, norm: $||v||$, normed vector space, rotational of vector function, scalar multiplication: $\alpha v$, Stokes' Theorem, surface integral, vector addition: $v + w$, vector Subtraction: $v - w$

## 1.1.   Normed Vector Space: $V(F)$

The term **normed vector space** [5, 6, 7] is used to name a mathematical structure where a norm [7] is defined as the rules in a non-empty set $V$ that meet the addition operation, **vector addition**, and the multiplication operation, **scalar multiplication**, between the elements of the set $V$ and the elements of a **field** $\mathbb{F}$. This normed vector space has two important operations **inner product** [8] and **cross product** [8].

**Definition 1.1.** A **normed vector space** $V$ over a **field** $\mathbb{F} \in \mathbb{R}^n$ is an algebraic structure where a set of elements called **vectors** $v, u, w \in V$ and a set of elements called **scalars** $\alpha, \beta \in \mathbb{F}$, together with two operations, **vector addition** and **scalar multiplication**, satisfy the next eight axioms [1, 4]:

Property 1. $u + (v + w) = (u + v) + w$
Property 2. $u + v = v + u$

Property 3. $\exists 0 \in V$ i called the zero vector, such that $\forall v \in V, v + 0 = v$

Property 4. $\forall v \in V, \exists -v \in V,$ such that $v + (-v) = 0$

Property 5. $\alpha, \beta \in \mathbb{F}, \alpha(\beta v) = (\alpha \beta) v$

Property 6. $1v = v$

Property 7. $\alpha(u + v) = \alpha u + \alpha v$

Property 8. $(\alpha + \beta) v = \alpha v + \beta v$

Remark 1.1. As it will be explained later in the chapter (Sect. 1.5), although the representation of the vectors can be in n-dimensional space, not all the operators act in this space [9, 10].

## 1.2. Basic operators

### 1.2.1. Vector Addition: $v + w$

There are two types of vectors, those that start anchored at the origin of the reference system **fixed vectors**, i.e. to a plane $\mathbb{R}^2$ or space $\mathbb{R}^3$, and those whose start is not anchored at the origin of the system **non-fixed vectors**.

**Definition 1.2.** The vector addition operation $\oplus : V \times V \to V$ takes two vectors $v \in \mathbb{R}^n$ and $w \in \mathbb{R}^n$, and assigns a third vector expressed as $v + w \in \mathbb{R}^n$.

Example 1.1. Let two vectors $v$ and $w \in \mathbb{R}^2$ be over the field $\mathbb{R}$, $v = (1, 2)$ and $w = (3, -1)$. What is $v + w$?

**Solution 1.1.** If $v = (v_1, v_2)$ and $w = (w_1, w_2) \Rightarrow v + w = (v_1 + w_1, v_2 + w_2)$, then $v + w = (4, 1)$.

Remark 1.2. The addition of two **fixed vectors** yields a **fixed vector**.

### 1.2.2. Vector Substraction: $v - w$

**Definition 1.3.** The vector substraction operation $\oplus : V \times V \to V$ takes two vectors $v \in \mathbb{R}^n$ and $w \in \mathbb{R}^n$, and assigns a third vector expressed as $v - w \in \mathbb{R}^n$ where $v - w \neq w - v$.

Example 1.2. Let two vectors $v$ and $w \in \mathbb{R}^3$ be over the field $\mathbb{R}$, $v = (1, 2, -1)$ and $w = (3, -1, 0)$. (i) What is $v - w$? (ii) What is $w - v$? (iii) Explain why $v - w \neq w - v$.

**Solution 1.2.** (i) If $v = (v_1, v_2, v_3)$ and $w = (w_1, w_2, w_3) \Rightarrow v - w = (v_1 - w_1, v_2 - w_2, v_3 - w_3)$, then $v - w = (1, 2, -1) - (3, -1, 0) = (-2, 3, -1)$. (ii) $w - v = (3, -1, 0) - (1, 2, -1) = (2, -3, 1)$. (iii) In general $v - w \neq w - v$ since $v_1 - w_1 \neq w_1 - v_1$, where $v_1, w_1$ are elements of the field $\mathbb{F}$.

Remark 1.3. Any non-fixed vector can be expressed as the subtraction of two **fixed vectors**.

The **addition** of the vectors $v + (-w)$ is equivalent to $v - w$, so this **vector addition** is known as **vector subtraction**.

Two vectors are **equal** if there is a translation between them. In this sense, a **fixed vector** and a **non-fixed vector** can be the same vector.

### 1.2.3.   Scalar Multiplication: $\alpha v$

**Definition 1.4.** The scalar multiplication operation $\otimes : \mathbb{F} \times V \rightarrow V$ takes any vector $v \in \mathbb{R}^n$ and a scalar $\alpha \in \mathbb{R}$, and assigns a third vector $\alpha v \in \mathbb{R}^n$, i.e. $\alpha v = \alpha(v_1, v_2, \cdots, v_n) = (\alpha v_1, \alpha v_2, \cdots, \alpha v_n)$. When the scalar $\alpha$ multiplies vector $v$, the length of vector $\alpha v$ will increase or decrease; however, if $\alpha = -1$ the vector $\alpha v$ keeps its length but not its orientation, which will be opposite.

Example 1.3. Given vector $v = (-3, 4, 5) \in \mathbb{R}^3$ and scalar $\alpha = -3 \in \mathbb{R}$, what is vector $\alpha v$?

**Solution 1.3.** $\alpha v = (-3)(-3, 4, 5) = (9, -12, -15)$.

This operation $\alpha v$ makes possible to increase the length of a vector (if $\alpha > 1$), decrease it (if $0 < \alpha < 1$), or change its orientation (if $\alpha < 0$).

### 1.2.4.   Norm: $||v||$

**Definition 1.5.** The **norm** (Eq. 1.1) of a **fixed vector** $a \in \mathbb{R}^n$ represents the length or **distance** with respect to point 0.

$$||a|| = \sqrt{\sum_{i=1}^{n} a_i^2}, \text{ where } a \in \mathbb{R}^n. \tag{1.1}$$

The **norm** (Eq. 1.1) of a **non-fixed vector** $c \in \mathbb{R}^n$ represents the length or **distance** (Eq. 1.2) between the **fixed vectors** $a, b \in \mathbb{R}^n$.

$$||c|| = ||a - b|| = \sqrt{\sum_{i=1}^{n} (a_i - b_i)^2}, \text{ where } c = a - b. \tag{1.2}$$

Example 1.4. There are two **fixed vectors** in a space $v = (3, 1, -2)$ and $w = (1, -1, 1)$. (i) What is the norm (or length) of vector $v$? (ii) What is the distance between the **fixed vectors** $v$ and $w$?

**Solution 1.4.** (i) The norm of vector $v$ is $||v|| = \sqrt{3^2 + 1^2 + (-2)^2} = \sqrt{14}$. (ii) The distance is $||v - w|| = \sqrt{(3-1)^2 + (1-(-1))^2 + ((-2)-1)^2} = \sqrt{17}$

It is important to differentiate the **norm** of a vector $||a||$ from the absolute value of a scalar $|x|$. The first one is a vector, the second one is a real number.

### 1.2.5.   Inner product: $v \cdot w$

**Definition 1.6.** The **inner product** is an algebraic operator that involves two vectors $a, b \in \mathbb{R}^n$ (Eq. 1.3) and the angle $\theta$ between them (Eq. 1.4).

$$a \cdot b = a_1 b_1 + a_2 b_2 + \cdots + a_n b_n \tag{1.3}$$

$$a \cdot b = ||a|| \, ||b|| \cos \theta \qquad (1.4)$$

From Def. 1.4 the inner product $\max\{a \cdot b\}$ occurs when $\cos \theta = 1$, i.e. $a||b$, whilst $\min\{a \cdot b\}$ occurs when $\cos \theta = 0$, i.e. $a \perp b$, regardless of the dimensional space in which the vectors $a$ and $b$ are.

The **inner product** holds the next five properties for any non-zero vectors $v, w, u \in \mathbb{R}^n$ and scalars $\alpha, \beta \in \mathbb{R}$:

Property 1. $v \cdot u = u \cdot v$

 Proof.

$$v \cdot u = v_1 u_1 + v_2 u_2 + \cdots + v_n u_n = u_1 v_1 + u_2 v_2 + \cdots + u_n v_n = u \cdot v \qquad (1.5)$$

$\square$

Property 2. $v \cdot (u + w) = (v \cdot u) + (v \cdot w)$

 Proof.

$$\begin{aligned} v \cdot (u + w) &= [v_1(u_1 + w_1) + v_2(u_2 + w_2) + \cdots + v_n(u_n + w_n)] \\ &= (v_1 u_1 + v_2 u_2 + \cdots + v_n u_n) + (v_1 w_1 + v_2 w_2 + \cdots + v_n w_n) \quad (1.6) \\ &= (v \cdot u) + (v \cdot w) \end{aligned}$$

$\square$

Property 3. $v \cdot (\alpha u + w) = \alpha(v \cdot u) + (v \cdot w)$

 Proof.

$$\begin{aligned} v \cdot (\alpha u + w) &= \alpha[v_1(u_1 + w_1) + v_2(u_2 + w_2) + v_n(u_n + w_n)] \\ &= v_1 u_1 + v_1 w_1 + v_2 u_2 + v_2 w_2 + \cdots + v_n u_n + v_n w_n] \quad (1.7) \\ &= \alpha(v \cdot u) + (v \cdot w) \end{aligned}$$

$\square$

Property 4. $\alpha v \cdot \beta w = \alpha \beta (v \cdot w)$

 Proof.

$$\begin{aligned} \alpha v \cdot \beta w &= \alpha v_1 \beta w_1 + \alpha v_2 \beta w_2 + \cdots + \alpha v_n \beta w_n \\ &= \alpha \beta (v_1 w_1 + v_2 w_2 + \cdots + v_n w_n) \quad (1.8) \\ &= \alpha \beta (v \cdot w) \end{aligned}$$

$\square$

Property 5. $v \perp u \Leftrightarrow v \cdot u = 0$

 Proof.

$$\text{From Def. 1.4, if } \theta = \frac{\pi}{2} \Rightarrow \cos \theta = 0, \text{ so } v \cdot u = 0 \qquad (1.9)$$

$\square$

Example 1.5. Given vectors $a = (1,0,0)$ and $b = (0,0,1) \in \mathbb{R}^3$. (i) What is the inner product? (ii) what is the angle between them?

**Solution 1.5.** (i) $a \cdot b = (1,0,0) \cdot (0,0,1) = 0$. (ii) $\theta = \cos^{-1}\left(\frac{a \cdot b}{||a|| \, ||b||}\right) = \cos^{-1}\left(\frac{0}{1}\right)$
$\Rightarrow \theta = \frac{\pi}{2}$ radians.

## 1.2.6.   Outer Product: $v \times w$

This operator is reviewed here on a plane and in space, and it will be studied in Chap 2.

**Definition 1.7.** The **cross product** (or **vector product**) of vectors $a$ and $b$, $a \times b$, is represented by the determinant (Eq. 1.10).

$$a \times b = \begin{vmatrix} \mathbf{i} & \mathbf{j} & \mathbf{k} \\ a_1 & a_2 & a_3 \\ b_1 & b_2 & b_3 \end{vmatrix} \tag{1.10}$$

The **cross product** on a plane or space represents the **normal vector** $\eta = a \times b$, vector $\eta$ is normal to the plane generated by the linear combination of vectors $a$ and $b$, i.e. $\alpha a + \beta b$.

The **cross product** holds the next six properties for the non-zero vectors $v, w, u \in \mathbb{R}^3$ and the scalars $\alpha, \beta \in \mathbb{R}$:

Remark 1.4. The demonstrations derive from the properties of the determinants.

Property 1. $v \times v = 0$

Proof.

$$v \times v = \begin{vmatrix} \mathbf{i} & \mathbf{j} & \mathbf{k} \\ v_1 & v_2 & v_3 \\ v_1 & v_2 & v_3 \end{vmatrix} = \widehat{0} \tag{1.11}$$

$\square$

Property 2. $v \times u = -(u \times v)$

Proof.

$$v \times u = \begin{vmatrix} \mathbf{i} & \mathbf{j} & \mathbf{k} \\ v_1 & v_2 & v_3 \\ u_1 & u_2 & u_3 \end{vmatrix} = -\begin{vmatrix} \mathbf{i} & \mathbf{j} & \mathbf{k} \\ u_1 & u_2 & u_3 \\ v_1 & v_2 & v_3 \end{vmatrix} = -(u \times v) \tag{1.12}$$

$\square$

Property 3. $v \times (u + w) = (v \times u) + (v \times w)$

Proof.

$$v \times (u+w) = \begin{vmatrix} \mathbf{i} & \mathbf{j} & \mathbf{k} \\ v_1 & v_2 & v_3 \\ u_1+w_1 & u_2+w_2 & u_3+w_3 \end{vmatrix}$$

$$= \begin{vmatrix} \mathbf{i} & \mathbf{j} & \mathbf{k} \\ v_1 & v_2 & v_3 \\ u_1 & u_2 & u_3 \end{vmatrix} + \begin{vmatrix} \mathbf{i} & \mathbf{j} & \mathbf{k} \\ v_1 & v_2 & v_3 \\ w_1 & w_2 & w_3 \end{vmatrix} \qquad (1.13)$$

$$= (v \times u) + (v \times w)$$

$\square$

Property 4. $v \times (\alpha u + w) = \alpha(v \times u) + (v \times w)$

Proof.

$$v \times (\alpha u + w) = \begin{vmatrix} \mathbf{i} & \mathbf{j} & \mathbf{k} \\ v_1 & v_2 & v_3 \\ \alpha u_1+w_1 & \alpha u_2+w_2 & \alpha u_3+w_3 \end{vmatrix}$$

$$= \begin{vmatrix} \mathbf{i} & \mathbf{j} & \mathbf{k} \\ v_1 & v_2 & v_3 \\ \alpha u_1 & \alpha u_2 & \alpha u_3 \end{vmatrix} + \begin{vmatrix} \mathbf{i} & \mathbf{j} & \mathbf{k} \\ v_1 & v_2 & v_3 \\ w_1 & w_2 & w_3 \end{vmatrix} \qquad (1.14)$$

$$= \alpha(v \times u) + (v \times w)$$

$\square$

Property 5. $\alpha v \times \beta w = \alpha\beta(v \times w)$

Proof.

$$\alpha v \times \beta w = \begin{vmatrix} \mathbf{i} & \mathbf{j} & \mathbf{k} \\ \alpha v_1 & \alpha v_2 & \alpha v_3 \\ \beta w_1 & \beta w_2 & \beta w_3 \end{vmatrix} = \alpha\beta(v \times w) \qquad (1.15)$$

$\square$

Property 6. $||a \times b|| = ||a|| \, ||b|| \sin\theta$

Proof.

$$\begin{aligned} ||a \times b||^2 &= ||a||^2 ||b||^2 - (a \cdot b)^2 \\ &= ||a||^2 ||b||^2 - ||a||^2 ||b||^2 \cos^2\theta \qquad (1.16) \\ &= ||a||^2 ||b||^2 \sin^2\theta \end{aligned}$$

$||a \times b|| = ||a|| \, ||b|| \, |\sin\theta|$ (Rmk. 1.5)                     $\square$

Remark 1.5. $\sqrt{\alpha^2} = |\alpha|$.

The geometrical representation of $||a \times b||$ is the area of a parallelogram with sides $a$ and $b$ that is the **norm** of vector $||\eta|| = ||a \times b||$ and its length. The expression $||a \times b||$ is equivalent to $||a \times b|| = ||a|| \, ||b|| \sin\theta$, where $\theta$ is the angle between vectors $a$ and $b$.

Example 1.6. Given **unit vectors** $i = (1,0,0)$ and $j = (0,1,0)$. (i) What is $i \times j$? (ii) What is $||i \times i||$? (iii) What is the angle between them? (iv) What is the area generated by these vectors?

**Solution 1.6.** (i)

$$(1,0,0) \times (0,1,0) = \begin{vmatrix} \mathbf{i} & \mathbf{j} & \mathbf{k} \\ 1 & 0 & 0 \\ 0 & 1 & 0 \end{vmatrix} = \mathbf{i} \begin{vmatrix} 0 & 0 \\ 1 & 0 \end{vmatrix} - \mathbf{j} \begin{vmatrix} 1 & 0 \\ 0 & 0 \end{vmatrix} + \mathbf{k} \begin{vmatrix} 1 & 0 \\ 0 & 1 \end{vmatrix} = (0,0,1)$$

(ii) $||a \times b|| = ||(0,0,1)|| = \sqrt{1^2} = 1$. (iii) $\theta = \sin^{-1} \dfrac{||a \times b||}{||a|| \, ||b||} = \dfrac{\pi}{2}$ radians. (iv) The area is represented by $||a \times b|| = ||(0,0,1)|| = 1$.

## 1.3.   Vector-Valued Functions

**Definition 1.8.** A vector-valued function is a transformation [1, 11, 12] $F : U \subset \mathbb{R}^n \to \mathbb{R}^n$ that assigns to each point $\mathbf{x} \in U$ a vector $F(\mathbf{x})$. In the case of the vector-valued function $F(x,y) = (F_1(x,y), F_2(x,y))$ the real-valued functions $F_1$ and $F_2$ are the components of the function $F$ and its **graph** is called **vector field** [13].

The procedure for plotting a vector $(x_0, y_0)$ in the Cartesian coordinate system is [1, 4]:

**Rule** 1. Plot the point $(x_0, y_0)$ in the Cartesian coordinate system.
**Rule** 2. Evaluate the point $(x_0, y_0)$ in the vector-valued function $F(x,y)$.
**Rule** 3. Move the Cartesian coordinate system to the point $(x_0, y_0)$.
**Rule** 4. Plot the vector $F(x_0, y_0)$ in the new Cartesian coordinate system.

Example 1.7. Let the function $F(x,y) = (3x, 5y)$. What is the vector associated with point $(1, 2)$?

**Solution 1.7.** The vector $c$ associated with point $(1, 2)$ is $(3, 10)$.

## 1.3.1.   Rotational of a Vector Function

The **rotational** operator of a vector-valued function $F$, **rot F**, is the measure of the rotation of the flow, i.e. the vectors that form a vector field around a particular point conforming a tiny paddle wheel at the point evaluated. There is a difference in the intensity of the vector field as it is represented by vectors of different lengths above and below the paddle wheel, thus inducing a **rotational movement** around the point [1, 14].

This operator is local, this means that its measure corresponds to a point in a plane or space and it can be featured by what happens in the neighborhood of the point. When this operator is part of the integrand of an integral (Sect. 1.4.1, 1.4.2), the rotation of the field in the region of the plane or space can be evaluated.

**Definition 1.9.** For a vector-valued function $F : \mathbb{R}^3 \to \mathbb{R}, (F_1(\mathbf{x}), F_2(\mathbf{x}), F_3(\mathbf{x}))$, where the real-valued functions $F_i : \mathbb{R}^n \to \mathbb{R}, i = 1, \cdots, 3$ are of class $C^1$, the rotational (Eq. 1.17) of the vector-valued function $F$ at point $\mathbf{x_0}$ is the **vector-valued function** defined by **rot F**.

$$\mathbf{rot\ F}\Big|_{\mathbf{x_0}} = \begin{pmatrix} \mathbf{i} & \mathbf{j} & \mathbf{k} \\ \dfrac{\partial}{\partial x} & \dfrac{\partial}{\partial y} & \dfrac{\partial}{\partial z} \\ F_1 & F_2 & F_3 \end{pmatrix}\Bigg|_{\mathbf{x_0}} \tag{1.17}$$

$$= \left(\frac{\partial F_3}{\partial y} - \frac{\partial F_2}{\partial z}\right)\mathbf{i} - \left(\frac{\partial F_3}{\partial x} - \frac{\partial F_1}{\partial z}\right)\mathbf{j} + \left(\frac{\partial F_2}{\partial x} - \frac{\partial F_1}{\partial y}\right)\mathbf{k}$$

Example 1.8. Let the vector-valued function $F : \mathbb{R}^3 \to \mathbb{R}, x^3 y + x^2 z^2 - \sin xyz$. (i) What is the **rot F**? (ii) What is the **rot F**$(0, 1, 2)$? (iii) Is the **rot F**$(0, 1, 2)$ truly representative of the rotational of $F$ on the open set $\mathbb{R}^3$?

**Solution 1.8.** (i)

$$\mathbf{rot\ F}\Big|_{\mathbf{x_0}} = \begin{pmatrix} \mathbf{i} & \mathbf{j} & \mathbf{k} \\ \dfrac{\partial}{\partial x} & \dfrac{\partial}{\partial y} & \dfrac{\partial}{\partial z} \\ x^3 y & x^2 z^2 & -\sin xyz \end{pmatrix}\Bigg|_{\mathbf{x_0}} = (-xz\cos xyz - 2x^2 z, \, yz\cos xyz, \, 2xz^2 - x^3)$$

(ii) **rot F**$(0, 1, 2) = (-xz\cos xyz - 2x^2 z, \, yz\cos xyz, \, 2xz^2 - x^3)\big|_{(0,1,2)} = 2$. (iii) No, it is not. The **rot F** operator is a **local** operator, so its value will depend on each point.

## 1.3.2.   Divergence of a Vector Function

The divergence operator of a vector-valued function $F$, $div F$, is a measure of the expansion or contraction of the vector field per unit of volume or area around the point $\mathbf{x_0}$.

**Definition 1.10.** For a vector-valued function $F : \mathbb{R}^n \to \mathbb{R}, (F_1(\mathbf{x}), F_2(\mathbf{x}), \cdots, F_n(\mathbf{x}))$, where the real-valued functions $F_i : \mathbb{R}^n \to \mathbb{R}, i = 1, \cdots, n$ are of class $C^1$, the divergence (Eq. 1.18) of the vector-valued function $F$ at point $\mathbf{x_0}$ is the **real-valued function** defined by

$$\mathbf{div\ F}\Big|_{\mathbf{x_0}} = \nabla \cdot F\Big|_{\mathbf{x_0}} = \frac{\partial F_1}{\partial x_1} + \frac{\partial F_2}{\partial x_2} + \cdots + \frac{\partial F_n}{\partial x_n} \tag{1.18}$$

This operator is local, which means that its measure corresponds to a point in the plane or space, therefore, it can only feature what happens in the neighborhood of a point. When this operator is in the integrand of an integral (Sect. 1.4.1, 1.4.2), the divergence of the field in the region of the plane or space can be evaluated.

Example 1.9. Let the vector-valued function $F : \mathbb{R}^3 \to \mathbb{R}, x^3y + x^2z^2 - \sin xyz$. (i) What is the **div F**? (ii) What is the **div** $\mathbf{F}(0,1,2)$? (iii) Is the **div** $\mathbf{F}(0,1,2)$ truly representative of the divergence of $F$ on the open set $\mathbb{R}^3$?

**Solution 1.9.** (i) The divergence of function $F$ is **div** $\mathbf{F} = \nabla \cdot F = \dfrac{\partial F_1}{\partial x_1} + \dfrac{\partial F_2}{\partial x_2} + \dfrac{\partial F_3}{\partial x_3}$
$= 3x^2y + 0 - xy\cos xyz$. (ii) **div** $\mathbf{F}(0,1,2) = 0$. (iii) No, it is not. The **div F** operator is a **local** operator, so its value will depend on each point.

## 1.3.3.   Line Integral

**Definition 1.11.** A Line Integral of a Vector Function measures the effect of a vector field $F : \mathbb{R}^n \to \mathbb{R}^n$ (Sect. 1.3) on an **oriented** trajectory $T : \mathbb{R} \to \mathbb{R}^n$ (Eq. 1.19). From the physical point of view, the line integral of a vector function measures the **work** done to move a particle on an oriented trajectory $T$ with the influence of a vector field $F$.

$$\oint_T F \circ T(t) \cdot T'(t)\, dt = \int_{t_1}^{t_2} F(T(t)) \cdot T'(t)\, dt, \text{ where } t \in [t_1, t_2] \subset \mathbb{R}. \quad (1.19)$$

Example 1.10. A particle moves along the oriented trajectory $T(t) = (t, t^4), t \in [0, \pi]$ and the force is represented by the vector field $F(x,y) = (-yx, \sin x)$. Compute the work done by the force field on a particle that moves along curve $T$.

**Solution 1.10.** $\oint_T F \circ T(t) \cdot T'(t)\, dt = \int_0^\pi (-t^5, \sin t) \cdot (1, 4t^3)\, dt = \int_0^\pi -t^5 + 4t^3$
$\sin t\, dt = -\dfrac{\pi^6 - 24\pi^3 + 144\pi}{6}$.

## 1.3.4.   Surface Integral

**Definition 1.12.** A Surface Integral of a Vector Function measures the effect of a vector field $F$ on an oriented region $S$ given by its normal vector $\eta(u,v)$ (Eq. 1.20). From the physical point of view, the surface integral of a vector function measures the **work** done over the oriented surface $S$ due to the influence of a vector field.

$$\oiint_S F \circ T(u,v) \cdot \eta(u,v)\, dS = \int_{u_1}^{u_2} \int_{v_1(u)}^{v_2(u)} F(T(u,v)) \cdot \frac{\partial T}{\partial u} \times \frac{\partial T}{\partial v}\, dv\, du. \quad (1.20)$$

Example 1.11. Let the external side surface of the open cylinder with radius 1 and height 2, and the vector field $F(x,y,z^2) = (x,y,z)$. Compute the surface integral of the vector function over the vector field.

**Solution 1.11.** Using $T(r,\theta) = (\cos\theta, \sin\theta, r)$ with $\theta \in [0, 2\pi], r \in [0,2]$.

$$\oiint_S F \circ T(r,\theta) \cdot \eta(r,\theta)\, dS = \int_{\theta_1}^{\theta_2} \int_{r_1}^{r_2} F(T(r,\theta)) \cdot \frac{\partial T}{\partial r} \times \frac{\partial T}{\partial \theta}\, dr\, d\theta$$

$$= \int_0^{2\pi} \int_0^2 F(T(r,\theta)) \cdot \frac{\partial T}{\partial r} \times \frac{\partial T}{\partial \theta}\, dr\, d\theta$$

$$= \int_0^{2\pi} \int_0^2 (\cos\theta, \sin\theta, r^2) \cdot (\cos\theta, \sin\theta, r)\, dr\, d\theta$$

$$= 12\pi \qquad (Note.\ 1.1).$$

$$(1.21)$$

Note 1.1. Note that the sign of the normal vector was changed.

## 1.4.   Vector Theorems

## 1.4.1.   Green's Theorem

**Definition 1.13.** Let $D \subset \mathbb{R}^2$ be a region and $\partial D$ be its closed counter-clockwise orientation boundary. Let the vector-valued function $F : \mathbb{R}^2 \to \mathbb{R}^2$ be on the $D$ region and the **unit vector k** [1, 4, 14, 15], then

$$\oint_{\partial D} F \circ c(t) \cdot c'(t)\, ds = \iint_D (\nabla \times F) \cdot \mathbf{k}\, dA.$$

Remark 1.6. Green's theorem states that the effect of the vector-valued function $F$ over the oriented closed curve $\partial D$, counter-clockwise orientation, is equivalent to the rotational effect over the area bounded by the region $D$.

Example 1.12. Let the vector-valued function $F(x,y) = (2x, 3y)$, the region $D$ bounded by $x \geq 0$, $y \geq 0$, $y = x^2$, and $y = x$. Verify Green's theorem.

**Solution 1.12.** $\oint_{\partial D} F \cdot ds = \oint_{C_1} F \times c_1(t) \cdot c_1'(t)\, dt + \oint_{C_2} F \times c_2(t) \cdot c_2'(t)\, dt$, where the curves $c_1(t) = (t, t^2), t \in [0,1]$, and $c_2(t) = (t,t), t \in [1,0]$ bound region $D$. $\int_0^1 (2t, 3t) \cdot (1,1)\, dt = \int_0^1 2t + 3t\, dt = \frac{5}{2}$. $\int_1^0 (2t, 3t^2) \cdot (1, 2t)\, dt = \int_1^0 2t + 6t^3\, dt = -\frac{5}{2}$, then $\oint_{\partial D} F \cdot ds = 0$. $\iint_D (\nabla \times F) \cdot \mathbf{k}\, dA = \int_0^1 \int_{x^2}^x (0,0,0) \cdot (0,0,1)\, dy\, dx = 0$. So, Green's theorem is verified.

## 1.4.2.   Stokes' Theorem

**Definition 1.14.** Let $D \subset \mathbb{R}^3$ be a closed region and $\partial D$ be its counter-clockwise orientation surface. Let the vector-valued function $F : \mathbb{R}^3 \to \mathbb{R}^3$ be on the $D$ region and the **normal vector** $T_v \times T_u$ [1, 4, 14, 15] be perpendicular to the surface $\partial D$.

$$\oint_{\partial D} F \circ c(t) \cdot c'(t)\, ds = \iint_S (\nabla \times F) \cdot T_v \times T_u\, dv\, du.$$

Remark 1.7. Stokes' theorem states that the effect of the vector-valued function $F$ over the oriented closed curve $\partial D$, counter-clockwise orientation, is equivalent to the rotational effect over the area bounded by the region $D$.

Example 1.13. Let the vector-value function $F(x,y) = (y,x,z)$, the region $D$ bounded by the cylinder surface $x^2 + y^2 = 1$, and the real-valued function $f(x,y) = 1 - x - y$. Verify Stokes' theorem.

**Solution 1.13.** $\oint_{\partial D} F(c(t)) \cdot c'(t) dt$, where the curve $c(t)$ is $c(t) = (\cos t, \sin t, 1 - \cos t - \sin t), t \in [0, 2\pi]$. $\int_0^{2\pi} (\sin t, \cos t, 1 - \cos t - \sin t) \cdot (-\sin t, \cos t, \sin t - \cos t)$

$dt = \int_0^{2\pi} -\cos^2 t \sin t + \sin^2 t \cos t + (1 - \cos t - \sin t)^2 (\sin t - \cos t) dt = 0$. On the

other hand $\iint_D \nabla \times F(T(r, \theta)) \cdot T_r \times T_\theta \, d\theta \, dr$, with $T(r, \theta) = (r\cos\theta, r\sin\theta, 1 -$

$\cos\theta - \sin\theta)$. $\int_0^1 \int_0^{2\pi} (0,0,0) \cdot (r\sin\theta - r\cos\theta, -\cos\theta - \sin\theta, \sin\theta + r\sin\theta) d\theta$

$dr = 0$.

## 1.4.3. Gauss' Theorem

**Definition 1.15.** Let $W \subset \mathbb{R}^3$ be a closed and solid region and $\partial W$ be its counter-clockwise orientation surface. Let the vector-valued function $F : \mathbb{R}^3 \to \mathbb{R}^3$ be on the $W$ region and the **normal vector** $T_v \times T_u$ [1, 4, 14, 15] be perpendicular to surface $W$.

$$\oiint_{\partial W} F \circ T(u,v) \cdot T_v \times T_u \, dv \, du = \iiint_W \nabla \cdot F \, dz \, dy \, dx.$$

Remark 1.8. Gauss' theorem states that the effect of the vector-valued function $F$ over the closed and oriented surface $\partial W$, counter-clockwise orientation, is equivalent to the divergent effect over the volume bounded by the region $W$.

Example 1.14. Let the vector-value function $F(x,y,z) = (x,y,z)$ and the region $W$ bounded by the solid cylinder $x^2 + y^2 = 1, 0 \le z \le 2$. Verify Gauss' theorem.

**Solution 1.14.** (i) $\iiint_{\partial W} \nabla \cdot F \, dz \, dy \, dx = \int_{-1}^1 \int_{-\sqrt{1-x^2}}^{\sqrt{1-x^2}} \int_0^2 3 \, dz \, dy$

$dx = \int_0^1 \int_0^{2\pi} \int_0^2 3r \, dz \, dr \, d\theta = 6\pi$. (ii) Surface integrals: (i.a) $\iint_W F(T(r,\theta)) \cdot T_\theta$

$\times T_r \, d\theta \, dr$, with $T(r,\theta) = (\cos\theta, \sin\theta, r) \Rightarrow \int_0^1 \int_0^{2\pi} (\cos\theta, \sin\theta, r) \cdot (\cos\theta, \sin\theta$

$,0) d\theta \, dr = 2\pi$. (i.b) $\iint_W F(T(r,\theta)) \cdot T_r \times T_\theta \, d\theta \, dr$, with $T(r,\theta) = (r\cos\theta, r\sin\theta$

$,0) \Rightarrow \int_0^1 \int_0^{2\pi} (r\cos\theta, r\sin\theta, 0) \cdot (0, 0, -r) \, d\theta \, dr = 0$. (i.c) $\iint_W F(T(r,\theta)) \cdot T_r \times$

$T_\theta \, dr \, d\theta$, with $T(r,\theta) = (r\cos\theta, r\sin\theta, 2) \Rightarrow \int_0^1 \int_0^{2\pi} (r\cos\theta, r\sin\theta, 2) \cdot (0,0,r)$

$d\theta \, dr = 2\pi$.

# 1.5.   Remarks

The cross product of the **Heaviside-Gibbs algebra** is frequently used to identify the sides of a surface, i.e. one side of the plane will be associated to the normal vector and the opposite side will be associated with the negative of the normal vector (Fig. 1.1).

Generally, to calculate the area of a surface it is not important what side of the area you use, nevertheless, there are cases where this is not so. To estimate the effect of a vector function $F$ over a surface, it is imperative to know what side of the surface will be used to calculate the effect of this vector function. The cross product is not defined in higher dimensions [8, 9, 10, 16, 17], the reason for this is structural, the cross product is essentially a determinant. In the case of two vectors in 4-dimensional space, the determinant is not defined and this substantially undermines the use of this algebra restricting it only to a plane or space.

The following chapters will be focussed on the **Grassmann algebra** [Die lineare Ausdehnungslehre, ein neuer Zweig der Mathematik, Hermann Grassmann, 1842] that acts on an $n$-dimensional space, as well as presenting the equivalent of the **Heaviside-Gibbs algebra**.

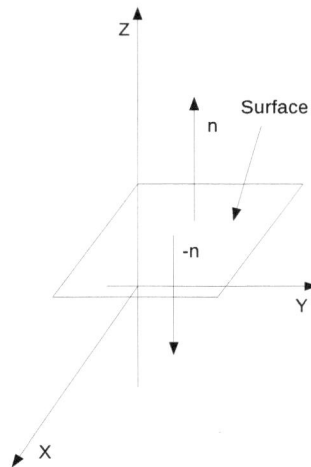

**Figure 1.1** Graphical interpretation of normal vector over a surface.

It is important that the reader reviews all the exercises in this chapter using the **Heaviside-Gibbs** algebra and solves them so in the following chapters, when the **Grassmann algebra** is introduced, he can have a full comprehension of both algebras and can compare them.

## 1.6.   Exercises

**Exercise 1.1.** Consider a segment of length $4\pi$ in $\mathbb{R}$. (i) Define a map from $\mathbb{R}$ to $\mathbb{R}^2$ that transforms this line into the ellipse $\frac{x^2}{a^2} + \frac{y^2}{b^2} = 1$ in the plane. (ii) Draw the figure. (iii) Discuss the nature of the parameterization.

**Exercise 1.2.** Consider a segment of length $3\pi$ in $\mathbb{R}$. Define a map from $\mathbb{R}$ to $\mathbb{R}^3$ that transforms this line into a unit circle projected onto the surface $f(x,y) = 1 - x - y^3$.

**Exercise 1.3.** Consider a unit circle in $\mathbb{R}^2$. Define a map from $\mathbb{R}^2$ to $\mathbb{R}^3$ that transforms the unit circle into the a paraboloid with concavity down and maximum in 1.

**Exercise 1.4.** Consider the mapping $T(t) = (\cos t, \sin t), t \in [0, 2\pi]$, and the vector field $F(x,y) = (-y^2, x)$. Compute the line integral of vector field in counter-clockwise direction.

**Exercise 1.5.** Consider a unit circle in the plane $f(x,y) = 2$ with centre at the origin and the vector field $F(x,y,z) = (-y^2, x, 1)$. (i) Compute the line integral of vector field in counter-clockwise direction.

**Exercise 1.6.** Consider a unit upper sphere and the vector field $F(x,y,z) = (y,x,z)$. (i) Compute the surface integral of vector field in counter-clockwise direction. (ii) Compute the line integral of vector field in clockwise direction.

**Exercise 1.7.** Consider the map $T(r,\theta) = (r\cos\theta, r\sin\theta)$, where $r \in [0,1]$ and $\theta \in [0, 2\pi]$. Compute the area of the parameterized surface.

**Exercise 1.8.** Let the vector-value function $F(x,y) = (2xy - x^2, x + y^2)$ [18] and the region $S$ (positive orientation) be bounded by $y = x^2$ and $y^2 = x$. Verify the Green' theorem.

**Exercise 1.9.** Let the vector-value function $F(x,y,z) = (3y, -xz, y^2)$ [18] and the region $S$ (positive orientation) be bounded by the paraboloid $2z = x^2 + y^2$, and by $z = 2$ and $C$. Verify the Stokes' theorem.

**Exercise 1.10.** Use the Gauss' theorem to calculate $\oiint_S F \, ds$, where $S$ is the lateral surface of the box $B$ with vertices $(\pm 1, \pm 2, \pm 3)$ and the normal vector $F(x,y,z) = (x^2 z^3, 2xyz^3, xz^4)$ [19] is pointing outwards.

# Part II

# Grassmann Algebra

**Geometric algebra** or **Grassmann algebra** is the central subject of this book. It has nine chapters: chapters 2 and 3 define this algebra in 2 and 3 dimensions; chapter 4 studies the extension to $n$ dimensions; in chapters 5 and 6 we reformulate the derivative and integral operators; from chapter 7 to 9 we focus on the **Geometric algebra** applications to introduce the Green's, Stokes', and Gauss' theorems in Differential forms; finally, in chapter 10 we see the **Fundamental Calculus Theorem** in terms of **Geometric algebra** and Differential forms.

# CHAPTER 2

# Geometric Algebra on $\mathbb{G}_2$

**Carlos Polanco**

Faculty of Sciences, Universidad Nacional Autónoma de México, México

**Abstract** This chapter is a review of **Geometric algebra** or **Grassmann algebra** on $\mathbb{G}_2$. This algebra is attributed to Hermann Grassmann [Die lineare Ausdehnungslehre, ein neuer Zweig der Mathematik 1842]. It has two main operators: **outer product** and **inner product**. Here, we will also study dot product, and geometric product, as well as their properties. We will start with the definition of Geometric algebra, its properties and most useful tools. With this background, we will define the differential forms in Chap. 5.

**Keywords**: Associativity: $a(bc) = (ab)c$, bivector, blades $< a >_i$, distributivity: $a(b+c)$, distributivity: $a \wedge (b+c)$, dual $Ia_r = b_{n-r}$, equation of a line, outer product, geometric algebra, geometric product, inner product, lines, multiplicative inverse: $a^{-1}$, norm $||a||$, reflections, reversion: $a^\dagger$, rotations

## 2.1.   Geometric Algebra on $\mathbb{G}_2$

**Definition 2.1.** The **Geometric algebra** or **Grassmann algebra** [1, 9, 20] is a unitary associative algebra, in symbols $\mathbb{G}_2 = \mathbb{G}_2(\mathbb{R}^2)$. It is formed by three elements: $\alpha$, **scalars**, $\sigma_1, \sigma_2$ **vectors**, and the elements $\sigma_1 \wedge \sigma_2$ named **bivectors**, or **equivalently** $\sigma_1 \sigma_2$, where $\alpha \in \mathbb{R}$. These elements will be expressed in **orthonormal** basis for convenience and they meet Eq. 2.1 for $i = j$.

$$\sigma_i \sigma_i = 1$$
$$\sigma_i \sigma_j = -\sigma_j \sigma_i \qquad (2.1)$$

An arbitrary element will be Eq. 2.2.

$$v = \underbrace{v_0}_{basis\,scalar} + \underbrace{v_1\sigma_1 + v_2\sigma_2}_{basis\,vector} + \underbrace{v_{12}\sigma_1 \wedge \sigma_2}_{basis\,bivector} \text{ in } \mathbb{G}_2. \tag{2.2}$$

Remark 2.1. An equivalent would be $\sigma_i \wedge \sigma_j$, $\sigma_i\sigma_j$, and $\sigma_{ij}$.

Example 2.1. Provide some examples of elements on $\mathbb{G}_2$.

**Solution 2.1.** $v = 4\sigma_2 + 5\sigma_1 \wedge \sigma_2$, $v = 4 + \sigma_2 + -4\sigma_{12}$, $v = -1 + \sigma_1 - 3\sigma_2 + 7\sigma_{12}$.

## 2.1.1.   Outer Product: $a \wedge b$

**Definition 2.2.** For two vectors $a = a_0 + a_1\sigma_1 + a_2\sigma_2 + a_3\sigma_{12}$ and $b = b_0 + b_1\sigma_1 + b_2\sigma_2 + b_3\sigma_{12} \in \mathbb{G}_2$ [1, 4, 8], we define

$$a \wedge b = \frac{1}{2}(ab - ba)$$

Example 2.2. Let two elements $a = (1, -1)$ and $b = (3, 2) \in \mathbb{G}_2$. (i) Obtain the outer product $a \wedge b = \frac{1}{2}(ab - ba)$. (ii) Obtain the geometric product using Def. 2.3.

**Solution 2.2.** (i) From Ex. 2.8 $a \wedge b = \frac{1}{2}(ab - ba) = 5\sigma_1\sigma_2$. (ii) $ab = a \cdot b + a \wedge b = 1 + 5\sigma_{12}$. So $a \wedge b = 5\sigma_{12}$.

The collinearity of two vectors implies that its **outer product** is zero, i.e. $a \wedge b = 0 \Leftrightarrow a \parallel b$.

Example 2.3. Let two collinear vectors $a = \sigma_1 + \sigma_2$ and $b = 2\sigma_1 + 2\sigma_2$. Determine the outer product.

**Solution 2.3.** $ab = 8$ and $ba = 8$, $a \wedge b = \frac{1}{2}(ab - ba) = 0$, so $a \parallel b$.

Example 2.4. Let the vectors $a = \sigma_1 + \sigma_{12}$ and $b = -2\sigma_1 + -3\sigma_2$. Determine the outer product.

**Solution 2.4.** $ab = -2 - 3\sigma_{12} + 2\sigma_2 - 3\sigma_1$ and $ba = -2 + 3\sigma_1 - 2\sigma_2 + 3\sigma_{12}$, $a \wedge b = \frac{1}{2}(ab - ba) = -2\sigma_2 + 3\sigma_1 + 3\sigma_{12}$.

## 2.1.2.   Inner Product: $a \cdot b$

**Definition 2.3.** For two elements $a = a_0 + a_1\sigma_1 + a_2\sigma_2 + a_3\sigma_{12}$ and $b = b_0 + b_1\sigma_1 + b_2\sigma_2 + b_3\sigma_{12} \in \mathbb{G}_2$ [1, 4, 8, 15], we define

$$a \cdot b = \frac{1}{2}(ab + ba)$$

Example 2.5. Consider elements $a = \sigma_1 + \sigma_2, b = \sigma_1 - \sigma_2 \in \mathbb{G}_2$ [1, 4, 15]. (i) Obtain the geometric products $ab$ and $ba$. (ii) Determine $a \cdot b$. (iii) Determine $a \wedge b$.

**Solution 2.5.** (i) $ab = -2, ba = 2\sigma_{12}$. (ii) $a \cdot b = 0$. (iii) $a \wedge b = 0$.

The perpendicularity of two vectors in $\mathbb{R}^2$ implies that the inner product is zero, i.e. $a \cdot b = 0 \Leftrightarrow a \perp b$.

Example 2.6. Let two perpendicular vectors [1, 21] $a = \sigma_1 + \sigma_2$ and $b = \sigma_1 - \sigma_2$ in $\mathbb{R}^2$. (i) Determine the inner product. (ii) Interpret geometrically the **inner product**.

**Solution 2.6.** (i) $ab = -2\sigma_{12}$ and $ba = 2\sigma_{12}$, $a \cdot b = \dfrac{1}{2}(ab + ba) = 0$, so $a \perp b$. (ii) See (Fig. 2.1).

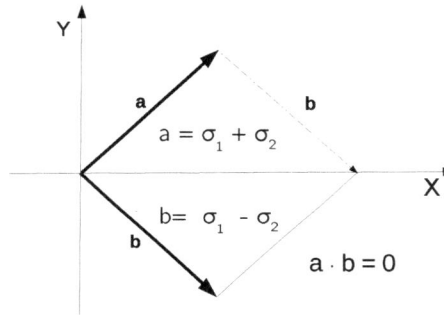

**Figure 2.1** Geometrical representation of $a \cdot b$

Example 2.7. Let the vectors $a = \sigma_1 + \sigma_{12}$ and $b = -2\sigma_1 + -3\sigma_2$. Determine the inner product.

**Solution 2.7.** $ab = -2 - 3\sigma_{12} + 2\sigma_2 - 3\sigma_1$ and $ba = -2 + 3\sigma_1 - 2\sigma_2 + 3\sigma_{12}$, $a \cdot b = \dfrac{1}{2}(ab + ba) = -2$.

## 2.1.3. Geometric Product: $ab$

From these two elements $a = a_0 + a_1\sigma_1 + a_2\sigma_2 + a_3\sigma_{12}$ and $b = b_0 + b_1\sigma_1 + b_2\sigma_2 + b_3\sigma_{12} \in \mathbb{G}_2$ [1, 2, 9, 10, 16, 17, 22, 23], the **geometric product** (Eq. 2.3) is defined as

$$\begin{aligned} \mathbf{ab} &= (a_0 + a_1\sigma_1 + a_2\sigma_2 + a_{12}\sigma_1\sigma_2)(b_0 + b_1\sigma_1 + b_2\sigma_2 + b_{12}\sigma_1\sigma_2) \\ &= a \cdot b + a \wedge b \end{aligned} \tag{2.3}$$

The geometric interpretation of the **bivector** $\sigma_1 \wedge \sigma_2$ is the **oriented area** with two sides A and B spanned by the vectors $\sigma_1$ and $\sigma_2$, whose value is 1 (Fig. 2.2). Similarly, $\sigma_1 \wedge -\sigma_2$ (Fig. 2.3) represents the area of side B and $-\sigma_1 \wedge -\sigma_2$ represents the area of side A.

Remark 2.2. Note that the vectorial expression implicitly defines the orientation of the surface.

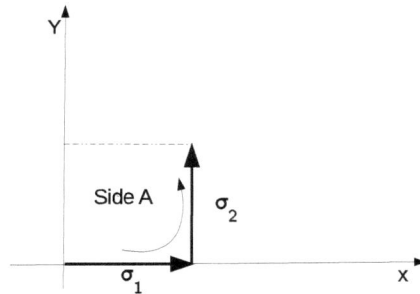

**Figure 2.2** Side A with area 1, $\sigma_1 \wedge \sigma_2$.

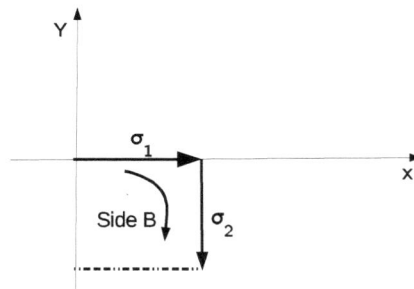

**Figure 2.3** Side B with area 1, $\sigma_1 \wedge -\sigma_2$.

Remark 2.3. When the elements on $\mathbb{G}_2$ are of the form $a = a_1\sigma_1 + a_2\sigma_2$ and $b = b_1\sigma_1 + b_2\sigma_2$, the **inner product** will only have the **scalar part and the outer product the vectorial part.**

Example 2.8. Let two elements $a = (1, -1)$ and $b = (3, 2) \in \mathbb{G}_2$ [1]. Express these elements using the orthonormal basis.

**Solution 2.8.** $a = \sigma_1 - \sigma_2$ and $b = 3\sigma_1 + 2\sigma_2$.

Example 2.9. Let the vectors $a = \sigma_1 + \sigma_{12}$ and $b = -2\sigma_1 + -3\sigma_2$. Express the geometric product $ab$ using the outer product and the inner product.

**Solution 2.9.** $ab = a \cdot b + a \wedge b = (-2) + (-2\sigma_2 + 3\sigma_1 + 3\sigma_{12})$.

For two elements $a = a_0 + a_1\sigma_1 + a_2\sigma_2 + a_3\sigma_{12}$ and $b = b_0 + b_1\sigma_1 + b_2\sigma_2 + b_3\sigma_{12} \in \mathbb{G}_2$ hold:

$$ab = (a_0 + a_1\sigma_1 + a_2\sigma_2 + a_3\sigma_{12})(b_0 + b_1\sigma_1 + b_2\sigma_2 + b_3\sigma_{12})$$
$$= (a_0 \quad (b_0 + b_1\sigma_1 + b_2\sigma_2 + b_3\sigma_{12}))$$
$$+ (a_1\sigma_1 (b_0 + b_1\sigma_1 + b_2\sigma_2 + b_3\sigma_{12}))$$
$$+ (a_2\sigma_2 (b_0 + b_1\sigma_1 + b_2\sigma_2 + b_3\sigma_{12}))$$
$$+ (a_3\sigma_{12}(b_0 + b_1\sigma_1 + b_2\sigma_2 + b_3\sigma_{12}))$$

$$= (a_0 b_0 + a_1 b_1 + a_2 b_2 - a_3 b_3)$$
$$+ (a_0 b_1 + a_1 b_0 - a_2 b_3 + a_3 b_2)\sigma_1$$
$$+ (a_0 b_2 + a_1 b_3 + a_2 b_0 - a_3 b_1)\sigma_2$$
$$+ (a_0 b_3 + a_1 b_2 - a_2 b_1 + a_3 b_0)\sigma_{12}$$

$$(2.4)$$

$$ba = (b_0 + b_1\sigma_1 + b_2\sigma_2 + b_3\sigma_{12})(a_0 + a_1\sigma_1 + a_2\sigma_2 + a_3\sigma_{12})$$
$$= (b_0 \quad (a_0 + a_1\sigma_1 + a_2\sigma_2 + a_3\sigma_{12}))$$
$$+ (b_1\sigma_1 (a_0 + a_1\sigma_1 + a_2\sigma_2 + a_3\sigma_{12}))$$
$$+ (b_2\sigma_2 (a_0 + a_1\sigma_1 + a_2\sigma_2 + a_3\sigma_{12}))$$
$$+ (b_3\sigma_{12}(a_0 + a_1\sigma_1 + a_2\sigma_2 + a_3\sigma_{12}))$$

$$= (a_0 b_0 + b_1 a_1 + a_2 b_2 - b_3 a_3)$$
$$+ (b_0 a_1 + b_1 a_0 - b_2 a_3 + b_3 a_2)\sigma_1$$
$$+ (b_0 a_2 + b_1 a_3 + b_2 a_0 - b_3 a_1)\sigma_2$$
$$+ (b_0 a_3 + b_1 a_2 - b_2 a_1 + b_3 a_0)\sigma_{12}$$

Remark 2.4. Verify that the inner product is $a \cdot b = (a_0 b_0 + b_1 a_1 + a_2 b_2 - b_3 a_3) + (b_0 a_1 + b_1 a_0)\sigma_1 + (a_0 b_2 + b_2 a_0)\sigma_2 + (a_3 b_0 + b_3 a_0)\sigma_{12}$ and the outer product is $a \wedge b = (b_2 a_3 - b_3 a_2)\sigma_1 + (a_0 b_2 + a_1 b_3 - a_3 b_1)\sigma_2 + (a_1 b_2 - a_2 b_1)\sigma_{12}$. Observe that $ab = a \cdot b + a \wedge b$.

From Eq. 2.4 the geometric product $ab$ on $\mathbb{G}_2$ has an element of the general form $\alpha \in \mathbb{R}$ (Eq. 2.5)

$$\alpha_0 + \alpha_1\sigma_1 + \alpha_2\sigma_2 + \alpha_{12}\sigma_{12} \qquad (2.5)$$

For two elements $a = a_1\sigma_1 + a_2\sigma_2$ and $b = b_1\sigma_1 + b_2\sigma_2 \in \mathbb{G}_2$ hold:

$$
\begin{aligned}
ab &= (a_1\sigma_1 + a_2\sigma_2)(b_1\sigma_1 + b_2\sigma_2) \\
&= (a_1\sigma_1(b_1\sigma_1 + b_2\sigma_2)) \\
&\quad + (a_2\sigma_2(b_1\sigma_1 + b_2\sigma_2)) \\[6pt]
&= (a_1b_1 + a_2b_2) \\
&\quad + (a_1b_2 - a_2b_1)\sigma_{12}
\end{aligned}
$$

$$
\begin{aligned}
ba &= (b_1\sigma_1 + b_2\sigma_2)(a_0 + a_1\sigma_1 + a_2\sigma_2) \\
&= (b_1\sigma_1(a_1\sigma_1 + a_2\sigma_2)) \\
&\quad + (b_2\sigma_2(a_1\sigma_1 + a_2\sigma_2)) \\[6pt]
&= (a_1b_1 + a_2b_2) \\
&\quad + (b_1a_2 - b_2a_1)\sigma_{12}
\end{aligned}
\tag{2.6}
$$

**Remark 2.5.** Verify that the inner product is $a \cdot b = a_1b_1 + a_2b_2$ and the outer product is $a \wedge B = (a_1b_2 - a_2b_1)\sigma_{12}$. Note that $ab = a \cdot b + a \wedge b$.

From Eq. 2.6, the geometric product $ab$ on $\mathbb{G}_2$ gives an element with the general form $\alpha \in \mathbb{R}$ (Eq. 2.7).

$$
\alpha_0 + \alpha_{12}\sigma_{12} \tag{2.7}
$$

For two elements $a = a_3\sigma_{12}$ and $b = b_3\sigma_{12} \in \mathbb{G}_2$ hold:

$$
\begin{aligned}
ab &= (a_3\sigma_{12})(b_3\sigma_{12}) \\
&= -(a_3b_3) \\[6pt]
ba &= (b_3\sigma_{12})(a_3\sigma_{12}) \\
&= -(b_3a_3)
\end{aligned}
\tag{2.8}
$$

**Remark 2.6.** Verify that the inner product is $a \cdot b = -(a_3b_3)$ and the outer product is $a \wedge B = 0$.

From Eq. 2.8, the geometric product $ab$ on $\mathbb{G}_2$ gives an element with the general form $\alpha \in \mathbb{R}$ (Eq. 2.9).

$$
\alpha_0 \tag{2.9}
$$

## 2.2.  Properties on $\mathbb{G}_2$

The following fundamental properties of geometric algebra [20, 23] will be exemplified, so it is advisable to check all the examples.

## 2.2.1.   Distributivity: $a(b+c)$

**Definition 2.4.** For three elements $a$, $b$, and $c \in \mathbb{G}_2$ [23, 24], then

$$a(b+c) = ab + ac$$

Proof.

$$
\begin{aligned}
ab + ac &= a \cdot b + a \wedge b + a \cdot c + a \wedge c \\
&= a \cdot b + a \cdot c + a \wedge b + a \wedge c \\
&= a \cdot (b+c) + a \wedge (b+c) \\
&= a(b+c)
\end{aligned}
\tag{2.10}
$$

$\square$

Example 2.10. Let three elements $a = \sigma_2, b = \sigma_1 - \sigma_2$, and $c = \sigma_1 + \sigma_2$. (i) Determine $a(b+c)$. (ii) Determine $ab + ac$. (iii) Is the distributivity fulfilled?

**Solution 2.10.** (i) $a(b+c) = \sigma_2(\sigma_1 - \sigma_2 + \sigma_1 + \sigma_2) = -2\sigma_{12}$. (ii) $ab = -\sigma_1\sigma_2 - 1$ and $ac = -\sigma_1\sigma_2 + 1$, then $ab + ac = -2\sigma_1\sigma_2$. (iii) From (i) and (ii) yes, it is.

## 2.2.2.   Distributivity: $a \wedge (b+c)$

**Definition 2.5.** For three elements $a$, $b$ and $c \in \mathbb{G}_2$ [1], hold:

$$a \wedge (b+c) = a \wedge b + a \wedge c$$

Proof.

$$
\begin{aligned}
a \wedge (b+c) &= -ab - ac + a \cdot (b+c) \\
&= -ab - ac + a \cdot b + a \cdot c \\
&= (-ab + a \cdot b) + (-ac + a \cdot c) \\
&= a \wedge b + a \wedge c
\end{aligned}
\tag{2.11}
$$

$\square$

Example 2.11. Let three elements $a = \sigma_2, b = \sigma_1 - \sigma_2$, and $c = \sigma_1 + \sigma_2$. (i) Determine $a \wedge (b+c)$. (ii) Determine $a \wedge b$ and $a \wedge c$. (iii) Is the distributivity over the outer product fulfilled?

**Solution 2.11.** (i) $a \wedge (b+c) = -2\sigma_{12}$. (ii) $a \wedge b = -\sigma_{12}$ and $a \wedge c = -\sigma_{12}$, then $a \wedge b + a \wedge c = -2\sigma_{12}$. (iii) Yes, it is.

Example 2.12. Let three elements $a = \sigma_1\sigma_2, b = 2 + \sigma_1 + \sigma_2$, and $c = -\sigma_1\sigma_2$ [1]. (i) Determine $a \wedge (b+c)$. (ii) Determine $a \wedge b$ and $a \wedge c$. (iii) Is the distributivity over the outer product fulfilled?

**Solution 2.12.** (i) $a \wedge (b+c) = \sigma_1 - \sigma_2$. (ii) $a \wedge b = -\sigma_1 + \sigma_2$ and $a \wedge c = 0$, then $a \wedge b + a \wedge c = \sigma_1 - \sigma_2$. (iii) Yes, it is.

## 2.2.3.  Multiplicative Inverse: $a^{-1}$

**Definition 2.6.** For an element $a \in \mathbb{G}_2$ [1], we define $a^{-1} = \dfrac{a}{a \cdot a}$ then

$$aa^{-1} = 1$$

Proof.

$$aa^{-1} = a\frac{a}{a \cdot a}$$

$$= \frac{aa}{a \cdot a}$$
$$= \frac{a \cdot a + a \wedge a}{a \cdot a} \qquad (2.12)$$
$$= \frac{a \cdot a}{a \cdot a} + \frac{a \wedge a}{a \cdot a}$$
$$= 1 + 0$$
$$= 1$$

$\square$

Example 2.13. Let element $a = 2\sigma_1 + \sigma_2 \in \mathbb{G}_2$. (i) Obtain the element $aa^{-1}$. (ii) Determine the geometric product $aa^{-1}$, the outer product $a \wedge a^{-1}$, and the inner product $a \cdot a^{-1}$.

**Solution 2.13.** (i) $a^{-1} = \dfrac{a}{5} = (\dfrac{2}{5}, \dfrac{1}{5}) \Rightarrow a^{-1} = \dfrac{2}{5}\sigma_1 + \dfrac{1}{5}\sigma_2$. (ii) $aa^{-1} = (2\sigma_1 + \sigma_2)(\dfrac{2}{5}\sigma_1 + \dfrac{1}{5}\sigma_2) = 1$. (iii) $a \cdot a^{-1} = 1$ and $a \wedge a^{-1} = 0$.

Example 2.14. Let element $a = 1 - \sigma_1 + 2\sigma_1\sigma_2 \in \mathbb{G}_2$. Obtain the elements $a^{-1}$ and $aa^{-1}$ [1].

**Solution 2.14.** $a^{-1} = \dfrac{a}{a \cdot a} = \dfrac{1 - \sigma_1 + 2\sigma_1\sigma_2}{-2 - 2\sigma_1 + 4\sigma_1\sigma_2} \Rightarrow aa^{-1} = \dfrac{(1 - \sigma_1 + 2\sigma_1\sigma_2)^2}{-2 - 2\sigma_1 + 4\sigma_1\sigma_2} = 1.$

## 2.2.4.  Associativity: $a(bc) = (ab)c$

**Definition 2.7.** For three elements $a$, $b$, and $c \in \mathbb{G}_2$ [1], then

$$(ab)c = a(bc)$$

Proof.

$$(ab)c = (a \cdot b + a \wedge b)c$$
$$= (a \cdot b)c + (a \wedge b)c$$
$$= a \cdot b \cdot c + (a \cdot b) \wedge c + (a \wedge b) \cdot c + a \wedge b \wedge c$$

$$(2.13)$$

$$a(bc) = a(b \cdot c + b \wedge c)$$
$$= a(b \cdot c) + a(b \wedge c)$$
$$= a \cdot b \cdot c + a \wedge (b \cdot c) + a \cdot (b \wedge c) + a \wedge b \wedge c$$

Note 2.1. $(a \cdot b) \wedge c + (a \wedge b) \cdot c = abc - cba$ and $a \wedge (b \cdot c) + a \cdot (b \wedge c) = abc - cba$.

Example 2.15. Consider three elements $i, j,$ and $\alpha j$, where $\alpha \in \mathbb{R}$ [1]. (i) Determine $i(j\alpha j)$. (ii) Determine $(ij)\alpha j$. (iii) Is the associativity fulfilled?

**Solution 2.15.** (i) $i(j\alpha j) = \sigma_1(\sigma_2 \alpha \sigma_2) = \alpha \sigma_1$. (ii) $(ij)\alpha j = (\sigma_1 \sigma_2)\alpha \sigma_2 = \alpha \sigma_1$. (iii) From the results (i) and (ii) yes, it is.

## 2.2.5.   Reversion: $a^\dagger$

**Definition 2.8.** For an element $a \in \mathbb{G}_2$ [1], we define the reversion of the element $a$ as $a^\dagger$, where
$$a^\dagger = (a_1 a_2 \cdots a_r)^\dagger = a_r \cdots a_2 a_1$$

Example 2.16. Let element $a = 1 - 3\sigma_1 - \sigma_1\sigma_2, \in \mathbb{G}_2$. Obtain the reverse of element $a$.

**Solution 2.16.** From the definition, the reversion of $a$ is $a^\dagger = 1 - 3\sigma_1 - \sigma_{21}$.

## 2.2.6.   Dual: $Ia_r$

**Definition 2.9.** Let two elements $a = a_0 + a_1\sigma_1 + a_2\sigma_2 + a_3\sigma_{12}$ and $b = b_0 + b_1\sigma_1 + b_2\sigma_2 + b_3\sigma_{12}$ both on $\mathbb{G}_2$, then

$$Ia_r = b_{n-r}$$

with $n = 2$

$$
\begin{array}{llll}
\text{scalar} & r = 0, & n - r = 2 & \text{bivector} \\
\text{vector} & r = 1, & n - r = 1 & \text{vector} \\
\text{bivector} & r = 2, & n - r = 0 & \text{scalar}
\end{array}
\tag{2.14}
$$

From Eq. 2.14
$$\text{scalar} \rightleftarrows \text{bivector}$$

Example 2.17. Explain the possibilities of the dual terms on $\mathbb{G}_2$ [17, 20, 23].

**Solution 2.17.** We have three alternatives: (i) If $r = 0$ and $n = 2$, then $Ia_0 = b_2$. (ii) If $r = 1$ and $n = 2$, then $Ia_1 = b_1$. (iii) If $r = 2$ and $n = 2$, then $Ia_2 = b_0$. From these three alternatives the only duality is the scalar $\rightleftarrows$ bivector.

Example 2.18. Let two elements $a = \sigma_1 + \sigma_2$ and $b = 2\sigma_1 + \sigma_2 \in \mathbb{G}_2$. Obtain the dual of the bivector $a \wedge b$.

**Solution 2.18.** $a \wedge b = -\sigma_{12}$, then $I(a \wedge b) = \sigma_{12}(-\sigma_{12}) = -\sigma_{12}\sigma_{12} = 1$.

### 2.2.7. Blades $< a >$

**Definition 2.10.** For an element $a = a_0 + a_1\sigma_1 + a_2\sigma_2 + a_3\sigma_{12}$ on $\mathbb{G}_2$, a vector space structure is Eq. 2.15 [17, 20, 23].

$$a = \underbrace{a_0}_{grade-0\,blade} + \underbrace{a_1\sigma_1 + a_2\sigma_2}_{grade-1\,blades} + \underbrace{a_3\sigma_{12}}_{grade-2\,blade} \tag{2.15}$$

where the element $a$ can be expressed as

$$a = \sum_{r=1}^{2} <a>_i \;=\; <a>_0 + <a>_1 + <a>_2$$

**Corollary 2.1.** Any element $a$ can be separated into the sum of even and odd blades.

$$< a > = < a >_+ + < a >_-$$

$$< a >_+ = < a >_0 + < a >_2 + \cdots + \tag{2.16}$$
$$< a >_- = < a >_1 + < a >_3 + \cdots +$$

Example 2.19. Let an element $a = 1 + 2\sigma_1 - 3\sigma_2 - 3\sigma_2\sigma_1, \in \mathbb{G}_2$. Obtain the blades of element $a$.

**Solution 2.19.** From the definition, $< a >_0 = 1$, $< a >_1 = 2\sigma_1 - 3\sigma_2$ and $< a >_2 = -3\sigma_2\sigma_1$.

### 2.2.8. Norm: $||a||$

**Definition 2.11.** Let the element $a = a_0 + a_1\sigma_1 + a_2\sigma_2 + a_3\sigma_{12}$ on $\mathbb{G}_2$, then its **norm** is the scalar part of Eq. 2.17 [17, 20, 23].

$$||a|| = \sqrt{aa^\dagger} \tag{2.17}$$

Example 2.20. Let an element $a = -1 + 2\sigma_1 - 3\sigma_2 - 3\sigma_2\sigma_1 \in \mathbb{G}_2$. Obtain the norm of element $a$.

**Solution 2.20.** From the definition, the norm of $a$ is Eq. 2.18

$$
\begin{aligned}
||a|| &= \sqrt{aa^\dagger} \\
&= \sqrt{(-1 + 2\sigma_1 - 3\sigma_2 - 3\sigma_{21})(-1 + 2\sigma_1 - 3\sigma_2 + 3\sigma_{21})} \\
&= \sqrt{23 + \cancel{14\sigma_1} \cancel{+ 18\sigma_2} \cancel{- \sigma_{12}}} \\
&= \sqrt{23}.
\end{aligned}
\tag{2.18}
$$

## 2.2.9.   Vector Components: $v_\parallel$ $v_\perp$

The components of a vector with the form $a = a_1\sigma_1 + a_2\sigma_2$, $v = v_1\sigma_1 + v_2\sigma_2$ on $\mathbb{G}_2$ are the orthonormal projection of the vector $v$ onto vector $a$ $v_\parallel$ and the orthonormal projection of vector $v$ onto vector $a$ $v_\perp$ (Fig. 2.4), such that vector $v = v_\parallel + v_\perp$ [25].

$$v_\parallel = \frac{v \cdot a}{||a||^2}a \tag{2.19}$$

$$v_\perp = v - v_\parallel \tag{2.20}$$

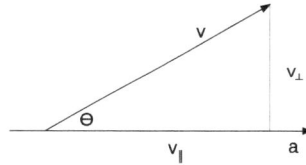

**Figure 2.4** Components of vector $v$.

Example 2.21. Let vectors $a = 2\sigma_2$ and $v = \sigma_1 + 2\sigma_2$ on $G_2$. Determine $v_\parallel$ and $v_\perp$.

**Solution 2.21.** $v \cdot a = 2$, then $v_\parallel = \frac{v \cdot a}{||a||^2}a = \frac{2}{2}2\sigma_2 = 2\sigma_2$. $v_\perp = v - v_\parallel = \sigma_1 + 2\sigma_2 - 2\sigma_2 = \sigma_1$.

## 2.3.   Reflections and Rotations on a Plane

Given a **vector** $a = a_1\sigma_1 + a_2\sigma_2 \in \mathbb{G}_2$, we define a **reflection** of vector $a$ as $Ia$ or $IIa$, where $I = \sigma_1 \wedge \sigma_2$ [1, 2, 9, 20, 22, 23, 24, 26]. A **rotation** is defined here as two successive reflections.

Example 2.22. Let an element $a = \sigma_1 + \sigma_2 \in \mathbb{G}_2$ [1]. (i) Determine $Ia$. (ii) Determine $aI$. (iii) Explain (i) and (ii).

**Solution 2.22.** (i) $Ia = \sigma_1\sigma_2 a = -\sigma_2 + \sigma_1 = \sigma_1 - \sigma_2$. (ii) $aI = a\sigma_1\sigma_2 = \sigma_2 - \sigma_1$. (iii) From these results, (i) is the **rotation** of $\frac{\pi}{2}$ in the **clockwise** direction and (ii) is the **rotation** of $\frac{\pi}{2}$ in the **counter-clockwise** direction.

Example 2.23. Let an element $a = \sigma_1 + \sigma_2 \in \mathbb{G}_2$ [1]. (i) Determine $IIa$. (ii) Determine $aII$. (iii) Explain (i) and (ii).

**Solution 2.23.** (i) $IIa = \sigma_1\sigma_2\sigma_1\sigma_2 a = -\sigma_1 - \sigma_2$. (ii) $aII = a\sigma_1\sigma_2\sigma_1\sigma_2 = -\sigma_1 - \sigma_2$. (iii) From these results, (i) is the **reflection** of $\pi$ in the **clockwise** direction and (ii) is the **reflection** of $\pi$ in the **counter-clockwise** direction.

Other method is the following: ( Case taken with the permission of the author [2]). To find the vector $y = -uxu$, where $u = \dfrac{u}{||u||}$ is the reflected vector and $x$ is the transformed vector., rotate through angle $\theta$ in the plane and perform two reflections in succession along any of the two axes in the plane, as long as (a) the angle between the axes is $\dfrac{\theta}{2}$ and (b) the rotation from the first axis to the second is in the same direction as the rotation to be performed (Figs. 2.5-2.7).

So if we want to rotate vector $v$, then $I = \sigma_{12}$. Let $m$ and $n$ be vectors along the axes satisfying the conditions, then the result of the rotation is

$$
\begin{aligned}
v' &= -m(-nvn^{-1})m^{-1} \\
&= (mn)v(mn)^{-1} \\
&= uvu^{-1}
\end{aligned}
\tag{2.21}
$$

where $u = mn$

The angle between vectors $v$ and $v''$ is $\theta$, twice the angle between $n$ and $m$ regardless of the value of $\varphi$ (Figs. 2.5, 2.6, and 2.5).

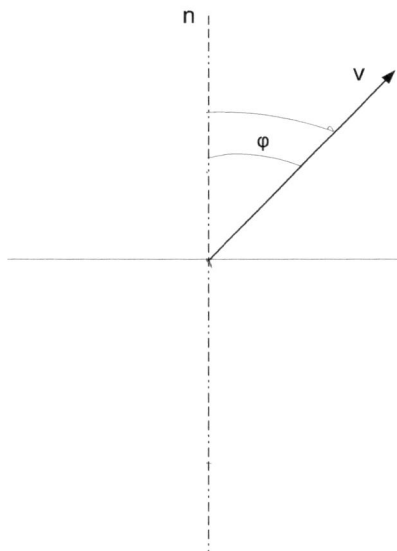

**Figure 2.5** First, the vector $v$ makes an angle $\varphi$ with axis $n$.

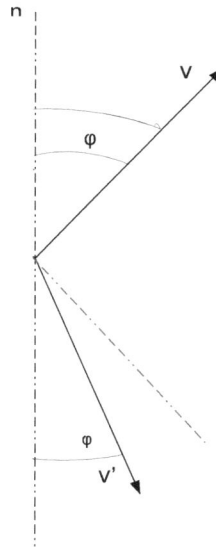

**Figure 2.6**  Second, the vector $v$ is then reflected along $n$ producing vector $v'$.

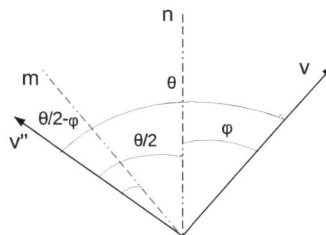

**Figure 2.7**  Finally, then along axis $m$ producing vector $v''$.

Example 2.24. Find the vector $y$ for the vector $x = \sigma_1 + 3\sigma_2$ over the vector $u = \sigma_1 + \sigma_2$.

**Solution 2.24.** $u = \dfrac{u}{||u||} = \dfrac{u}{2}$, then $y = -uxu = -\dfrac{3}{2}\sigma_1 - \dfrac{1}{2}\sigma_2$.

## 2.4.   Geometric Representation of a Line on $\mathbb{R}^2$

**Definition 2.12.** Given a **vector** $v$ and a point $x_0$ in space $\mathbb{R}^2$, what is the equation of the line passing through point $x_0$ in the direction of vector $v$ (Fig. 2.8)? [2, 9,

20, 22, 24] (Example adapted with permission of the author [22] and reproduced here from [1]).

The line $L_{x_0}(v)$ is given by

$$L_{x_0}(v) := \{\mathbf{x} \,|\, (\mathbf{x} - x_0) \wedge v = 0\} \tag{2.22}$$

Remark 2.7. Take note that the **oriented line** is defined by $v$.

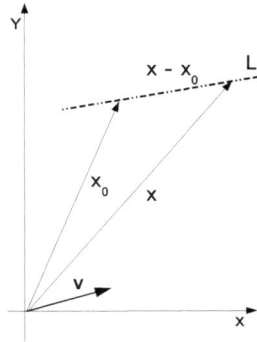

**Figure 2.8** The line $L_{x_0}$ through the point $x_0$ is in the direction of $v$. (Figure taken with the permission of the author [1]

Example 2.25. Given a vector $v$ and a point $x_0$ in space $\mathbb{R}^2$, what is the equation of the line passing through point $x_0 = (1,2)$ in the direction of vector $v = (1,1)$?

**Solution 2.25.** The line $L_{x_0}(v)$ is given by

$$L_{x_0=(1,2)}(v) := \{\mathbf{x} \,|\, (\mathbf{x} - x_0) \wedge v = 0\} \tag{2.23}$$

$$\begin{aligned}
[(x_1\sigma_1 + x_2\sigma_2) - (\sigma_1 + 2\sigma_2)] \wedge (\sigma_1 + \sigma_2) = 0 \\
[(x_1 - 1)\sigma_1 + (x_2 - 2)\sigma_2] \wedge (\sigma_1 + \sigma_2) = 0
\end{aligned} \tag{2.24}$$

The outer product $(x - x_0) \wedge v = \frac{1}{2}[(x - x_0)v - v(x - x_0)]$,

$$[(x_1 - 1)\sigma_1 + (x_2 - 2)\sigma_2](\sigma_1 + \sigma_2) = (x_1 + x_2 - 3) + (x_1 - x_2 + 1)\sigma_1\sigma_2 \tag{2.25}$$

$$(\sigma_1 + \sigma_2)[(x_1 - 1)\sigma_1 + (x_2 - 2)\sigma_2] = (x_1 + x_2 - 3) + (x_2 - x_1 - 1)\sigma_1\sigma_2$$

From Eqs. 2.25, $(x - x_0) \wedge v = x_1 - x_2 + 1 = 0 \Rightarrow x_1 = x_2 - 1$, the points $(x_2 - 1, x_2)$ are the solution. Note that the point $(1,2)$ meets the line $L_{x_0}(v)$.

## 2.5. Geometric Representation of a Plane on $\mathbb{R}^2$

**Definition 2.13.** Let two **vectors** $v$ and $u$, and a point $x_0$ on $\mathbb{R}^2$. What is the equation of the plane passing through point $x_0$ over the plane generated by the vectors $v$ and $u$? (Fig. 2.9) (Case adapted with permission of the author [1, 22]). The plane $P_{x_0}(u \wedge v)$ is given by

$$P_{x_0}(u \wedge v) := \{\mathbf{x} \,|\, (\mathbf{x} - x_0) \wedge (u \wedge v) = 0\} \tag{2.26}$$

Remark 2.8. Note that the **oriented area** in the plane is defined by $u \wedge v$.

$$\text{Plane} := (x\text{-}x_0) \wedge (u \wedge v) = 0$$

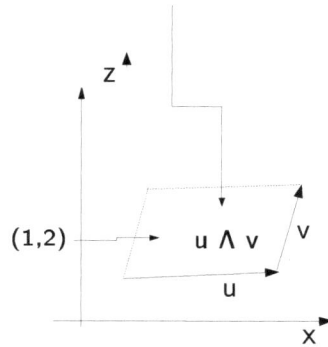

**Figure 2.9** The plane $P_{x_0}$ passing through point $x_0$ in the direction of the bivector $u \wedge v$.

Example 2.26. Let two vectors $v$ and $u$, and a point $x_0$ in space $\mathbb{R}^2$. What is the equation of the plane passing through point $x_0 = (1,2)$ in the plane generated by the vectors $v = (1,0)$ and $u = (0,1)$?

**Solution 2.26.** The plane $P_{x_0}(u, v)$ is given by

$$P_{x_0=(1,2)}(u \wedge v) := \{\mathbf{x} \,|\, (\mathbf{x} - x_0) \wedge (u \wedge v) = 0\} \tag{2.27}$$

$$\begin{aligned} [(x_1 \sigma_1 + x_2 \sigma_2) - (\sigma_1 + 2\sigma_2)] \wedge (\sigma_1 \sigma_2) = 0 \\ [(x_1 - 1)\sigma_1 + (x_2 - 2)\sigma_2] \wedge (\sigma_1 \sigma_2) = 0 \end{aligned} \tag{2.28}$$

From Eq. 4.13, $[(x_1 - 1)\sigma_1 + (x_2 - 2)\sigma_2][\sigma_1 \sigma_2] - [\sigma_1 \sigma_2][(x_1 - 1)\sigma_1 + (x_2 - 2)\sigma_2] = 0$. So, the points $(x_2 - 2, x_1 - 1)$ are the solution. Note that the point $(1,2)$ meets the plane $P_{x_0=(1,2)}(u \wedge v)$.

## 2.6.  Remarks

Note that the outer product $a \wedge b$ is neither scalar nor vector, in fact, it is a plane spanned by the vectors in the plane $\mathbb{R}^2$. $a$ and $b$ represent planes with an area and an orientation, if you interchange $a$ and $b$ you will change the sign of $a \wedge b$) [2, 9, 20, 22, 23, 24, 26]. This operator makes unnecessary the **cross product** used in **Heaviside-Gibbs algebra** to identify the sides of an area in the plane $\mathbb{R}^2$, that is the reason why in the first chapter we will only use the **cross product**. In this chapter we introduced the equivalent of the two algebras in the plane $\mathbb{R}^2$. In the next chapter, we will introduce the **Geometric algebra** or **Grassmann algebra** in the space $\mathbb{R}^3$.

As you will see in the next chapters, **Geometric algebra** acts in a space of $n$ dimensions, which makes possible to extend the orientation of objects to these dimensions.

We strongly advise the reader to review and solve all the exercises in this chapter, to get familiar with this algebra.

## 2.7.   Exercises

**Exercise 2.1.** Provide some exercises of elements on $\mathbb{G}_2$.

**Exercise 2.2.** Let two elements $a = (-1,1)$ and $b = (2,3) \in \mathbb{G}_2$ [1]. Express these elements using the orthonormal basis.

**Exercise 2.3.** Let two elements $a = \sigma_{12}$ and $b = \sigma_1 + \sigma_2 \in \mathbb{G}_2$. Obtain the exterior product $a \wedge b$.

**Exercise 2.4.** Let the vectors $a = -\sigma_1 + 2\sigma_2$ and $b = \sigma_1 + \sigma_2 \in \mathbb{G}_2$. Determine the exterior product.

**Exercise 2.5.** Let elements $a = \sigma_1, b = -\sigma_{21} \in \mathbb{G}_2$. (i) Obtain the geometric products $ab$ and $ba$. (ii) Determine $a \cdot b$. (iii) Determine $a \wedge b$.

**Exercise 2.6.** Let three elements $a = \sigma_1, b = \sigma_1 - \sigma_2$, and $c = \sigma_1 + \sigma_2 \in \mathbb{G}_2$ [1]. (i) Determine $a(b+c)$. (ii) Determine $ab + ac$. (iii) Is the distributivity fulfilled?

**Exercise 2.7.** Let three elements $a = \sigma_{12}, b = 3 + \sigma_1 - \sigma_2$, and $c = -\sigma_{12} \in \mathbb{G}_2$. (i) Determine $a(b+c)$ and $(b+c)a$. (ii) Determine $a \wedge (b+c)$. (iii) Determine $a \wedge b$ and $a \wedge c$. (iv) Is the distributivity over the exterior product fulfilled?

**Exercise 2.8.** Let an element $a = 2 - \sigma_1 + 2\sigma_1\sigma_2 \in \mathbb{G}_2$. Obtain the element $a^{-1}$.

**Exercise 2.9.** Let an element $a = \sigma_1\sigma_2 \in \mathbb{G}_2$. Obtain the reverse of element $a$.

**Exercise 2.10.** Let an element $a = 1 + 2\sigma_1\sigma_2 \in \mathbb{G}_2$. Obtain the blades.

**Exercise 2.11.** Let two elements $a = \sigma_1\sigma_2$ and $b = 2\sigma_1 + \sigma_2 \in \mathbb{G}_2$. Obtain the dual of $a \wedge b$.

**Exercise 2.12.** Let an element $a = 1 + \sigma_1 + \sigma_2 + \sigma_2\sigma_1 \in \mathbb{G}_2$. Obtain the norm of element $a$.

**Exercise 2.13.** Let three elements $a = \sigma_1, b = \sigma_2$, and $c = \alpha\sigma_1 - \sigma_2 \in \mathbb{G}_2$ where $\alpha \in \mathbb{R}$. (i) Determine $a(bc)$. (ii) Determine $(ab)c$. (iii) Is the associativity fulfilled?

**Exercise 2.14.** Let an element $a = \sigma_1 + \sigma_2 \in \mathbb{G}_2$ [1]. (i) Determine $Ia$. (ii) Determine $aI$. (iii) Explain (i) and (ii).

**Exercise 2.15.** Let an element $a = 2\sigma_1 + 3\sigma_2 \in \mathbb{G}_2$ [1]. (i) Determine $IIa$. (ii) Determine $aII$. (iii) Explain (i) and (ii).

**Exercise 2.16.** Find vector $y$ for the vector $x = 2\sigma_1 + \sigma_2$ onto vector $u = \sigma_1 - 2\sigma_2$.

**Exercise 2.17.** Let vector $v$ and a point $x_0$ in the line $\mathbb{R}^2$. What is the equation of the **line** passing through the point $x_0 = (1,1)$ in the direction of the vector $v = (1,0)$?

**Exercise 2.18.** Let two vectors $v$ and $u$, and a point $x_0$ in the $\mathbb{R}^2$ space. What is the equation of the **plane** passing through point $x_0 = (2,1)$ in the plane generated by the vectors $v = (0,1)$ and $u = (1,1)$?

**CHAPTER 3**

# Geometric Algebra on $\mathbb{G}_3$

**Carlos Polanco**

Faculty of Sciences, Universidad Nacional Autónoma de México, México

**Abstract** This chapter reviews and elaborates on the operators from Geometric algebra on $\mathbb{G}_2$ to $\mathbb{G}_3$. This algebra is attributed to Hermann Grassmann [Die lineare Ausdehnungslehre, ein neuer Zweig der Mathematik 1842]. It is formed by two main operators, the **outer product** and the **inner product**, it also includes the element called **bivector**. Here, we review their properties and their application in space.

**Keywords**: Associativity: $a(bc) = (ab)c$, bivector: $a \wedge b$, blades $< a >$, component: $v_{\parallel}$, component: $v_{\perp}$, distributivity: $a(b+c)$, distributivity: $a \wedge (b+c)$, dual $Ia_r = b_{n-r}$, equation of a line, outer product, geometric algebra, geometric product, inner product, lines, multiplicative inverse: $a^{-1}$, norm $||a||$, reflections, reversion: $a^{\dagger}$, rotations

## 3.1. Geometric Algebra on $\mathbb{G}_3$

**Definition 3.1.** The **Geometric algebra** or **Grassmann algebra** [1, 9] is a unitary associative algebra, in symbols $\mathbb{G}_3 = \mathbb{G}_3(\mathbb{R}^3)$. It is formed by eight $2^3$ elements: $\alpha$, **scalars**, $\sigma_1, \sigma_2, \sigma_3$ **vectors**, $\sigma_1 \wedge \sigma_2$, $\sigma_1 \wedge \sigma_3$, $\sigma_2 \wedge \sigma_3$ **bivectors**, and $\sigma_1 \wedge \sigma_2 \wedge \sigma_3$ **trivectors** (or **equivalently** $\sigma_1 \sigma_2 \sigma_3$), where $\alpha \in \mathbb{R}$. For convenience these elements are expressed in **orthonormal** basis and they meet Eqs. 3.1 for $i = j$.

$$\begin{aligned} \sigma_i \sigma_i &= 1 \\ \sigma_i \sigma_j &= -\sigma_j \sigma_i \end{aligned} \tag{3.1}$$

An arbitrary element is Eq. 3.2.

$$v = \underbrace{v_0}_{basis\,scalars}$$

$$+ \underbrace{v_1\sigma_1 + v_2\sigma_2}_{basis\,vectors}$$

$$+ \underbrace{v_{12}\sigma_1 \wedge \sigma_2 + v_{23}\sigma_2 \wedge \sigma_3 + v_{31}\sigma_3 \wedge \sigma_1}_{basis\,bivectors} \qquad (3.2)$$

$$+ \underbrace{v_{123}\sigma_1 \wedge \sigma_2 \wedge \sigma_3}_{basis\,trivector} \text{ in } \mathbb{G}_3.$$

Remark 3.1. The equivalent would be $\sigma_i \wedge \sigma_j$, $\sigma_i\sigma_j$, and $\sigma_{ij}$.

Example 3.1. Provide some examples of elements on $\mathbb{G}_2$.

**Solution 3.1.** $v = 4\sigma_2 + 5\sigma_1 \wedge \sigma_2$, $v = 4 + \sigma_2 + -4\sigma_{12}$, $v = -1 + \sigma_1 - 3\sigma_2 + 7\sigma_{12}$.

### 3.1.1.   Outer Product: $a \wedge b$

**Definition 3.2.** For two elements $a = a_0 + a_1\sigma_1 + a_2\sigma_2 + a_{12}\sigma_1 \wedge \sigma_2 + a_{13}\sigma_1 \wedge \sigma_3 + a_{23}\sigma_2 \wedge \sigma_3 + \sigma_1 \wedge \sigma_2 \wedge \sigma_3$ and $b = b_0 + b_1\sigma_1 + b_2\sigma_2 + b_{12}\sigma_1 \wedge \sigma_2 + b_{13}\sigma_1 \wedge \sigma_3 + b_{23}\sigma_2 \wedge \sigma_3 + b_{123}\sigma_1 \wedge \sigma_2 \wedge \sigma_3 \in \mathbb{G}_3$ [1], we define

$$a \wedge b = \frac{1}{2}(ab - ba)$$

Remark 3.2. If $a \wedge b = 0 \Rightarrow a \parallel b$.

Example 3.2. Consider two elements $a = \sigma_{21} + \sigma_{123}$ and $b = 2 + \sigma_{12} \in \mathbb{G}_3$. Obtain the outer product $a \wedge b$.

**Solution 3.2.** Since $ab = -1 - \sigma_3 - 2\sigma_{12} + 2\sigma_{123}$ and $ba = 1 - \sigma_3 - 2\sigma12 + 2\sigma_{123}$, $a \wedge b = 0$.

Example 3.3. Are vectors $a = \sigma_1 + \sigma_2 + \sigma_3$ and $b = 2\sigma_1 + 2\sigma_2 + 2\sigma_3$ colinear?. (i) Determine the outer product. (ii) What about Eq. 3.2.

**Solution 3.3.** (i) $ab = 6$ and $ba = 6$, $a \wedge b = \frac{1}{2}(ab - ba) = 0$, so $a \parallel b$. (ii) Yes, both elements are parallel.

### 3.1.2.   Inner Product: $a \cdot b$

**Definition 3.3.** For two elements $a = a_0 + a_1\sigma_1 + a_2\sigma_2 + a_{12}\sigma_1 \wedge \sigma_2 + a_{13}\sigma_1 \wedge \sigma_3 + a_{23}\sigma_2 \wedge \sigma_3 + \sigma_1 \wedge \sigma_2 \wedge \sigma_3$ and $b = b_0 + b_1\sigma_1 + b_2\sigma_2 + b_{12}\sigma_1 \wedge \sigma_2 + b_{13}\sigma_1 \wedge \sigma_3 + b_{23}\sigma_2 \wedge \sigma_3 + b_{123}\sigma_1 \wedge \sigma_2 \wedge \sigma_3 \in \mathbb{G}_3$ [1], we define

$$a \cdot b = \frac{1}{2}(ab + ba).$$

Remark 3.3. If $a \cdot b = 0 \Leftrightarrow a \perp b$.

Example 3.4. Let two elements $a = \sigma_1\sigma_2\sigma_3, b = \sigma_1 - \sigma_2\sigma_1\sigma_3 \in \mathbb{G}_3$. (i) Obtain the geometric products $ab$ and $ba$. (ii) Determine $a \cdot b$. (iii) Determine $a \wedge b$.

**Solution 3.4.** (i) $ab = (\sigma_1\sigma_2\sigma_3)(\sigma_1 - \sigma_2\sigma_1\sigma_3) = \sigma_2\sigma_3 - 1$, $ba = \sigma_2\sigma_3 - 1$. (ii) $a \cdot b = \sigma_2\sigma_3 - 1$. (iii) $a \wedge b = 0$.

Example 3.5. Let two elements $a = \sigma_1\sigma_2$ and $b = \sigma_3 \in \mathbb{G}_3$. (i) Obtain the geometric products $ab$ and $ba$. (ii) From Def. 3.3 determine $a \cdot b$. (iii) From Def. 3.2 determine $a \wedge b$.

**Solution 3.5.** (i) $ab = \sigma_1\sigma_2\sigma_3$. $ba = \sigma_1\sigma_2\sigma_3$. (ii) $a \cdot b = \sigma_1\sigma_2\sigma_3$. (iii) $a \wedge b = 0$.

Example 3.6. Let two elements $a = 1 + \sigma_1 + \sigma_2 - \sigma_2\sigma_3$ and $b = \sigma_1\sigma_2 \in \mathbb{G}_3$. (i) Obtain the geometric products $ab$ and $ba$. (ii) From Def. 3.3 determine $a \cdot b$. (iii) From Def. 3.2 determine $a \wedge b$.

**Solution 3.6.** (i) $ab = (1 + \sigma_1 + \sigma_2 - \sigma_2\sigma_3)(\sigma_1\sigma_2) = \sigma_1 - \sigma_2 + \sigma_1\sigma_2 - \sigma_1\sigma_3$. $ba = (\sigma_1\sigma_2)(1 + \sigma_1 + \sigma_2 - \sigma_2\sigma_3) = -\sigma_1 + \sigma_2 + \sigma_1\sigma_2 - \sigma_1\sigma_3$. (ii) $a \cdot b = \sigma_1\sigma_2 - \sigma_1\sigma_3$. (iii) $a \wedge b = \sigma_1 - \sigma_2$.

Example 3.7. Let two elements $a$ and $b$ on $\mathbb{G}_3$ [1] $a = \sigma_1 + \sigma_2 + \sigma_3$ and $b = \sigma_1 + \sigma_2 - 2\sigma_3$ in $\mathbb{R}^3$. (i) Determine the inner product. (ii) Give a geometrical interpretation of the **inner product**.

**Solution 3.7.** (i) $ab = 3\sigma_{13} + 3\sigma_{23}$ and $ba = -3\sigma_{13} - 3\sigma_{23}$, $a \cdot b = \dfrac{1}{2}(ab + ba) = 0$, so $a \perp b$. (ii) See Fig. 3.1.

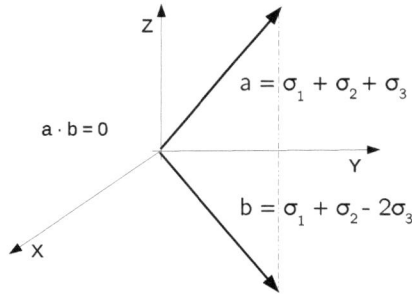

**Figure 3.1** Geometrical representation of $a \cdot b$.

### 3.1.3.　Geometric Product: $ab$

For two elements $a = a_0 + a_1\sigma_1 + a_2\sigma_2 + a_{12}\sigma_1 \wedge \sigma_2 + a_{13}\sigma_1 \wedge \sigma_3 + a_{23}\sigma_2 \wedge \sigma_3 + \sigma_1 \wedge \sigma_2 \wedge \sigma_3$ and $b = b_0 + b_1\sigma_1 + b_2\sigma_2 + b_{12}\sigma_1 \wedge \sigma_2 + b_{13}\sigma_1 \wedge \sigma_3 + b_{23}\sigma_2 \wedge \sigma_3 + b_{123}\sigma_1 \wedge \sigma_2 \wedge \sigma_3 \in \mathbb{G}_3$ [1], the **geometric product** (Eq. 3.3) is defined as

$$\mathbf{ab} = (a_0 + a_1\sigma_1 + a_2\sigma_2 + a_{12}\sigma_1\sigma_2 + a_{13}\sigma_1\sigma_3 + a_{23}\sigma_2\sigma_3 + a_{123}\sigma_1\sigma_2\sigma_3)$$
$$(b_0 + b_1\sigma_1 + b_2\sigma_2 + b_{12}\sigma_1\sigma_2 + b_{13}\sigma_1\sigma_3 + b_{23}\sigma_2\sigma_3 + b_{123}\sigma_1\sigma_2\sigma_3) \quad (3.3)$$
$$= a \cdot b + a \wedge b$$

Where the term $\mathbf{a} \cdot \mathbf{b}$ is called **inner product** (Def. 3.1.2) and the term $\mathbf{a} \wedge \mathbf{b}$ is called **outer product** (Def. 3.1.1) [1].

Remark 3.4. When the elements on $\mathbb{G}_3$ are of the form $a = a_1\sigma_1 + a_2\sigma_2 + a_3\sigma_3$ and $b = b_1\sigma_1 + b_2\sigma_2 + a_3\sigma_3$, the **inner product** only has the **scalar part** and the **outer product** the **vectorial part**.

Note 3.1. The geometrical description of the **trivector** $\sigma_1 \wedge \sigma_2 \wedge \sigma_3$ is the **oriented volume** spanned by the vectors $\sigma_1$, $\sigma_2$ and $\sigma_3$ (Fig. 3.2).

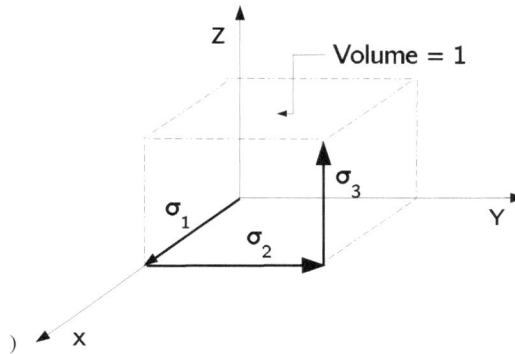

**Figure 3.2** The **oriented volume** formed by the vectors $\sigma_1$, $\sigma_2$ and $\sigma_3$. (Figure taken from [1])

Example 3.8. Let two elements $a = \sigma_1 - 2\sigma_2 + 3\sigma_3 - \sigma_1\sigma_2$ and $b = 2\sigma_1 + 3\sigma_2 + \sigma_3 \in \mathbb{G}_3$ [1]. (i) Obtain the geometric product $ab$ and $ba$. (ii) Obtain $a \wedge b = \frac{1}{2}(ab - ba)$. (iii) Obtain $a \cdot b = \frac{1}{2}(ab + ba)$. (iv) Explain the results obtained in (iii).

**Solution 3.8.** (i) $ab = (\sigma_1 - 2\sigma_2 + 3\sigma_3 - \sigma_1\sigma_2)(2\sigma_1 + 3\sigma_2 + \sigma_3) = -1 - 3\sigma_1 + 2\sigma_2 + 7\sigma_1\sigma_2 - 5\sigma_1\sigma_3 - 11\sigma_2\sigma_3 - \sigma_1\sigma_2\sigma_3$ $ba = (2\sigma_1 + 3\sigma_2 + \sigma_3)(\sigma_1 - 2\sigma_2 + 3\sigma_3 - \sigma_1\sigma_2) = -1 + 3\sigma_1 - 2\sigma_2 - 7\sigma_1\sigma_2 + 5\sigma_1\sigma_3 + 11\sigma_2\sigma_3 - \sigma_1\sigma_2\sigma_3$. (ii) $a \wedge b = -6\sigma_1 + 4\sigma_2 + 14\sigma_1\sigma_2 - 10\sigma_1\sigma_3 - 22\sigma_2\sigma_3$. (iii) $a \cdot b = -1 - \sigma_1\sigma_2\sigma_3$. (iv) The **geometric product** is the addition of the **inner product** and the **outer product**, but the **inner product** is **not** the real part of the geometric product.

For two elements $a = a_0 + a_1\sigma_1 + a_2\sigma_2 + a_3\sigma_3 + a_4\sigma_{12} + a_5\sigma_{23} + a_6\sigma_{31} + a_7\sigma_{123}$ and $b = b_0 + b_1\sigma_1 + b_2\sigma_2 + b_3\sigma_3 + b_4\sigma_{12} + b_5\sigma_{23} + b_6\sigma_{31} + b_7\sigma_{123} \in \mathbb{G}_3$, we observe

$$ab = (a_0 \quad (b_0 + b_1\sigma_1 + b_2\sigma_2 + b_3\sigma_3 + b_4\sigma_{12} + b_5\sigma_{23} + b_6\sigma_{31} + b_7\sigma_{123}))$$
$$+ (a_1\sigma_1 \ (b_0 + b_1\sigma_1 + b_2\sigma_2 + b_3\sigma_3 + b_4\sigma_{12} + b_5\sigma_{23} + b_6\sigma_{31} + b_7\sigma_{123}))$$
$$+ (a_2\sigma_2 \ (b_0 + b_1\sigma_1 + b_2\sigma_2 + b_3\sigma_3 + b_4\sigma_{12} + b_5\sigma_{23} + b_6\sigma_{31} + b_7\sigma_{123}))$$
$$+ (a_3\sigma_3 \ (b_0 + b_1\sigma_1 + b_2\sigma_2 + b_3\sigma_3 + b_4\sigma_{12} + b_5\sigma_{23} + b_6\sigma_{31} + b_7\sigma_{123}))$$
$$+ (a_4\sigma_{12} (b_0 + b_1\sigma_1 + b_2\sigma_2 + b_3\sigma_3 + b_4\sigma_{12} + b_5\sigma_{23} + b_6\sigma_{31} + b_7\sigma_{123}))$$
$$+ (a_5\sigma_{23} (b_0 + b_1\sigma_1 + b_2\sigma_2 + b_3\sigma_3 + b_4\sigma_{12} + b_5\sigma_{23} + b_6\sigma_{31} + b_7\sigma_{123}))$$
$$+ (a_6\sigma_{31} (b_0 + b_1\sigma_1 + b_2\sigma_2 + b_3\sigma_3 + b_4\sigma_{12} + b_5\sigma_{23} + b_6\sigma_{31} + b_7\sigma_{123}))$$
$$+ (a_7\sigma_{123}(b_0 + b_1\sigma_1 + b_2\sigma_2 + b_3\sigma_3 + b_4\sigma_{12} + b_5\sigma_{23} + b_6\sigma_{31} + b_7\sigma_{123}))$$

$$= (a_0b_0 + a_1b_1 + a_2b_2 + a_3b_3 - a_4b_4 - a_5b_5 - a_6b_6 - a_7b_7)$$
$$+ (a_0b_1 + a_1b_0 - a_2b_4 - a_3b_6 + a_4b_2 - a_5b_7 + a_6b_3 - a_7b_5)\sigma_1$$
$$+ (a_0b_2 + a_1b_4 + a_2b_0 - a_3b_5 - a_4b_1 + a_6b_7 + a_7b_3 + a_7b_6)\sigma_2$$
$$+ (a_0b_3 + a_1b_6 + a_2b_5 + a_3b_0 - a_4b_2 - a_5b_2 - a_6b_1 - a_7b_4)\sigma_3$$
$$+ (a_0b_4 + a_1b_2 - a_2b_1 + a_3b_7 + a_4b_0 + a_5b_6 - a_6b_5 + a_7b_3)\sigma_{12}$$
$$+ (a_0b_6 + a_1b_3 - a_2b_7 - a_3b_1 + a_4b_5 - a_5b_4 + a_6b_0 - a_7b_2)\sigma_{13}$$
$$+ (a_0b_5 + a_1b_7 + a_2b_3 - a_3b_2 - a_4b_6 + a_5b_0 + a_6b_4 + a_7b_1)\sigma_{23}$$
$$+ (a_0b_7 + a_1b_5 - a_2b_6 + a_3b_4 + a_4b_3 + a_5b_1 - a_6b_2 + a_7b_0)\sigma_{123}$$

$$ba = (b_0 \quad (a_0 + a_1\sigma_1 + a_2\sigma_2 + a_3\sigma_3 + a_4\sigma_{12} + a_5\sigma_{23} + a_6\sigma_{31} + a_7\sigma_{123}))$$
$$+ (b_1\sigma_1 \ (a_0 + a_1\sigma_1 + a_2\sigma_2 + a_3\sigma_3 + a_4\sigma_{12} + a_5\sigma_{23} + a_6\sigma_{31} + a_7\sigma_{123}))$$
$$+ (b_2\sigma_2 \ (a_0 + a_1\sigma_1 + a_2\sigma_2 + a_3\sigma_3 + a_4\sigma_{12} + a_5\sigma_{23} + a_6\sigma_{31} + a_7\sigma_{123}))$$
$$+ (b_3\sigma_3 \ (a_0 + a_1\sigma_1 + a_2\sigma_2 + a_3\sigma_3 + a_4\sigma_{12} + a_5\sigma_{23} + a_6\sigma_{31} + a_7\sigma_{123}))$$
$$+ (b_4\sigma_{12} (a_0 + a_1\sigma_1 + a_2\sigma_2 + a_3\sigma_3 + a_4\sigma_{12} + a_5\sigma_{23} + a_6\sigma_{31} + a_7\sigma_{123}))$$
$$+ (b_5\sigma_{23} (a_0 + a_1\sigma_1 + a_2\sigma_2 + a_3\sigma_3 + a_4\sigma_{12} + a_5\sigma_{23} + a_6\sigma_{31} + a_7\sigma_{123}))$$
$$+ (b_6\sigma_{31} (a_0 + a_1\sigma_1 + a_2\sigma_2 + a_3\sigma_3 + a_4\sigma_{12} + a_5\sigma_{23} + a_6\sigma_{31} + a_7\sigma_{123}))$$
$$+ (b_7\sigma_{123}(a_0 + a_1\sigma_1 + a_2\sigma_2 + a_3\sigma_3 + a_4\sigma_{12} + a_5\sigma_{23} + a_6\sigma_{31} + a_7\sigma_{123}))$$

$$= (a_0b_0 + a_1b_1 + a_2b_2 + a_3b_3 - a_4b_4 - a_5b_5 - a_6b_6 - a_7b_7)$$
$$+ (b_0a_1 + b_1a_0 - b_2a_4 - b_3a_6 + b_4a_2 - b_5a_7 + b_6a_3 - b_7a_5)\sigma_1$$
$$+ (b_0a_2 + b_1a_4 + b_2a_0 - b_3a_5 - b_4a_1 + b_5a_3 + b_6a_7 + b_7a_6)\sigma_2$$
$$+ (b_0a_3 + b_1a_6 + b_2a_5 + b_3a_0 - b_4a_7 - b_5a_2 - b_6a_1 - b_7a_4)\sigma_3$$
$$+ (b_0a_4 + b_1a_2 - b_2a_1 + b_3a_7 + b_4a_0 + b_5a_6 - b_6a_5 + b_7a_3)\sigma_{12}$$
$$+ (b_0a_6 + b_1a_3 - b_2a_7 - a_1b_3 + b_4a_5 - b_5a_4 + b_6a_0 - b_7a_2)\sigma_{13}$$
$$+ (b_0a_5 + b_1a_7 + b_2a_3 - b_3a_2 - b_4a_6 + b_5a_0 - b_6a_4 + b_7a_1)\sigma_{23}$$
$$+ (b_0a_7 + b_1a_5 - b_2a_6 + b_3a_4 + b_4a_3 + b_5a_1 - b_6a_2 + b_7a_0)\sigma_{123}$$

$$(3.4)$$

**Remark 3.5.** Verify that the inner product is $a \cdot b = (a_0b_0 + b_1a_1 + a_2b_2 - b_3a_3 - a_4b_4 - a_5b_5 - a_6b_6 - a_7b_7) + (a_0b_1 + a_1b_0 - a_5b_7 - a_7b_5)\sigma_1 + (a_0b_2 + a_2b_0 + a_6b_7 - a_4b_1)\sigma_2 + (a_0b_3 + a_3b_0 - a_4b_7 - a_7b_4)\sigma_3 + (a_0b_4 + a_3b_7 + a_4b_7 + a_7b_3)\sigma_{12} + (a_0b_6 - a_2b_7 + a_6b_0 - a_7b_2)\sigma_{13} + (a_0b_5 + a_1b_7 + a_5b_0 + a_7b_1)\sigma_{23} + (a_0b_7 - a_2b_6 +$

$a_4b_3 + a_7b_0)\sigma_{123}$ and the outer product is $a \wedge b = (-a_2b_4 - a_3b_6 + a_4b_2 + a_6b_3)\sigma_1 + (a_1b_4 - b_2a_3 - a_3b_5 - a_4b_1)\sigma_2 + (a_1b_6 + a_2b_5 - a_5b_2 - a_6b_1)\sigma_3 + (a_1b_2 - a_2b_1 + a_4b_0 + a_5b_6 - a_6b_5)\sigma_{12} + (a_1b_3 - a_2b_7 - a_3b_1 + a_4b_5 - a_5b_4 - a_7b_2)\sigma_{13} + (a_1b_7 + a_2b_3 - a_3b_2 - a_4b_6 + a_6b_4)\sigma_{23} + (a_1b_5 + a_3b_4 + a_5b_1 - a_6b_2)$
$\sigma_{123}$. Note that $ab = a \cdot b + a \wedge b$.

From Eq. 3.4, the Geometric product $ab$ on $\mathbb{G}_3$ produces an element with the general form of Eq. 3.5 $\alpha \in \mathbb{R}$.

$$\alpha_0 + \alpha_1\sigma_1 + \alpha_2\sigma_2 + \alpha_{12}\sigma_{12} + \alpha_{23}\sigma_{23} + \alpha_{13}\sigma_{13} + \alpha_{123}\sigma_{123} \qquad (3.5)$$

For two elements $a = a_1\sigma_1 + a_2\sigma_2 + a_3\sigma_3$ and $b = b_1\sigma_1 + b_2\sigma_2 + b_3\sigma_3 \in \mathbb{G}_3$, we observe

$$\begin{aligned}
ab &= (a_1\sigma_1 + a_2\sigma_2 + a_3\sigma_3)(b_1\sigma_1 + b_2\sigma_2 + b_3\sigma_3) \\
&= (a_1\sigma_1 (b_1\sigma_1 + b_2\sigma_2 + b_3\sigma_3)) \\
&+ (a_2\sigma_2(b_1\sigma_1 + b_2\sigma_2 + b_3\sigma_3)) \\
&+ (a_3\sigma_3(b_1\sigma_1 + b_2\sigma_2 + b_3\sigma_3)) \\[8pt]
&= (a_1b_1 + a_2b_2 + a_3b_3) \\
&+ (a_1b_2 - a_2b_1)\sigma_{12} \\
&+ (a_1b_3 - a_3b_1)\sigma_{13} \\
&+ (a_2b_3 - a_3b_2)\sigma_{32}
\end{aligned}$$

$$\begin{aligned}
ba &= (b_1\sigma_1 + b_2\sigma_2 + b_3\sigma_3)(a_0 + a_1\sigma_1 + a_2\sigma_2 + a_3\sigma_3) \\
&= (b_1\sigma_1 (a_1\sigma_1 + a_2\sigma_2 + a_3\sigma_3)) \\
&+ (b_2\sigma_2(a_1\sigma_1 + a_2\sigma_2 + a_3\sigma_3)) \\
&+ (b_3\sigma_3(a_1\sigma_1 + a_2\sigma_2 + a_3\sigma_3)) \\[8pt]
&= (a_1b_1 + a_2b_2 + a_3b_3) \\
&+ (b_1a_2 - b_2a_1)\sigma_{12} \\
&+ (b_1a_3 - b_3a_1)\sigma_{13} \\
&+ (b_2a_3 - b_3a_2)\sigma_{32}
\end{aligned}$$

$$(3.6)$$

Remark 3.6. Verify that the inner product is $a \cdot b = a_1b_1 + a_2b_2 + a_3b_3$ and the outer product is $a \wedge B = (b_1a_2 - a_2b_1)\sigma_{12} + (a_1b_3 - a_3b_1)\sigma_{13} + (a_2b_3 - a_3b_2)\sigma_{32}$. Notice that $ab = a \cdot b + a \wedge b$.

From Eq. 3.6 the Geometric product $ab$ on $\mathbb{G}_3$ produces an element with the general form of Eq. 3.7, $\alpha \in \mathbb{R}$.

$$\alpha_0 + \alpha_{12}\sigma_{12} + \alpha_{23}\sigma_{23} + \alpha_{13}\sigma_{13} \qquad (3.7)$$

For two elements $a = a_3\sigma_{123}$ and $b = b_3\sigma_{123} \in \mathbb{G}_3$ we observe

$$ab = (a_3 \sigma_{123})(b_3 \sigma_{123})$$
$$= -(a_3 b_3)$$

$$ba = (b_3 \sigma_{123})(a_3 \sigma_{123})$$
$$= -(b_3 a_3)$$

(3.8)

**Remark 3.7.** Verify that the inner product is $a \cdot b = -(b_3 a_3)$ and the outer product is $a \wedge B = 0$.

From Eq. 3.8, the Geometric product $ab$ on $\mathbb{G}_3$ produces an element with the general form of Eq. 3.9 $\alpha \in \mathbb{R}$.

$$\alpha_0$$

(3.9)

## 3.2.    Properties on $\mathbb{G}_3$

The following fundamental properties of Geometric algebra [20, 23] are exemplified, therefore, it is advisable to review these examples.

### 3.2.1.    Distributivity: $a(b+c)$

**Definition 3.4.** For three elements $a$, $b$, and $c \in \mathbb{G}_3$ [23, 24], then

$$a(b+c) = ab + ac$$

Proof.

$$ab + ac = a \cdot b + a \wedge b + a \cdot c + a \wedge c$$
$$= a \cdot b + a \cdot c + a \wedge b + a \wedge c$$
$$= a \cdot (b+c) + a \wedge (b+c)$$
$$= a(b+c)$$

(3.10)

□

Example 3.9. Let three elements $a = \sigma_3, b = \sigma_1 - \sigma_2,$ and $c = \sigma_1 + \sigma_2$. (i) Determine $a(b+c)$. (ii) Determine $ab + ac$. (iii) Is the distributivity fulfilled?

**Solution 3.9.** (i) $a(b+c) = \sigma_2(\sigma_1 - \sigma_2 + \sigma_1 + \sigma_2) = -2\sigma_{13}$. (ii) $ab = -\sigma_1\sigma_3 + \sigma_2\sigma_3$ and $ac = -\sigma_1\sigma_3 - \sigma_2\sigma_3$, then $ab + ac = -2\sigma_1\sigma_3$. (iii) From the results of (i) and (ii) yes, it is.

### 3.2.2.    Distributivity: $a \wedge (b+c)$

**Definition 3.5.** For three elements $a$, $b$, and $c \in \mathbb{G}_3$ [1], then

$$a \wedge (b+c) = a \wedge b + a \wedge c$$

Proof.

$$
\begin{aligned}
a \wedge (b+c) &= -ab - ac + a \cdot (b+c) \\
&= -ab - ac + a \cdot b + a \cdot c \\
&= (-ab + a \cdot b) + (-ac + a \cdot c) \\
&= a \wedge b + a \wedge c
\end{aligned}
\tag{3.11}
$$

$\square$

Example 3.10. Let three elements $a = 1 + \sigma_1, b = \sigma_1 \sigma_3,$ and $c = \sigma_1 + \sigma_2$. (i) Determine $a \wedge (b+c)$. (ii) Determine $a \wedge b + a \wedge c$. (iii) Is the distributivity fulfilled?

**Solution 3.10.** (i) $a \wedge (b+c) = \sigma_3 + \sigma_{12}$. (ii) $a \wedge b + a \wedge c = \sigma_3 + \sigma_{12}$. (iii) From the results of (i) and (ii) yes, it is.

### 3.2.3.    Multiplicative Inverse: $a^{-1}$

**Definition 3.6.** For an element $a \in \mathbb{G}_3$ [1], we define $a^{-1} = \dfrac{a}{a \cdot a}$, then

$$
aa^{-1} = 1
$$

Proof.

$$
\begin{aligned}
aa^{-1} &= a \frac{a}{a \cdot a} \\
\\
&= \frac{aa}{a \cdot a} \\
&= \frac{a \cdot a + a \wedge a}{a \cdot a} \\
&= \frac{a \cdot a}{a \cdot a} + \frac{a \wedge a}{a \cdot a} \\
&= 1 + 0 \\
&= 1
\end{aligned}
\tag{3.12}
$$

$\square$

Example 3.11. Let an element $a = 1 - \sigma_1 + 2\sigma_1 \sigma_3 \in \mathbb{G}_3$. Obtain the elements $a^{-1}$ and $aa^{-1}$ [1].

**Solution 3.11.** $a^{-1} = \dfrac{a}{a \cdot a} = \dfrac{1 - \sigma_1 + 2\sigma_1 \sigma_3}{-2 - 2\sigma_1 + 4\sigma_1 \sigma_3} \Rightarrow aa^{-1} = \dfrac{(1 - \sigma_1 + 2\sigma_1 \sigma_3)^2}{-2 - 2\sigma_1 + 4\sigma_1 \sigma_2}$
$= 1$.

### 3.2.4.    Associativity: $a(bc) = (ab)c$

**Definition 3.7.** For three elements $a$, $b$, and $c \in \mathbb{G}_3$ [1], then

$$
(ab)c = a(bc)
$$

Proof.

$$(ab)c = (a \cdot b + a \wedge b)c$$
$$= (a \cdot b)c + (a \wedge b)c$$
$$= a \cdot b \cdot c + (a \cdot b) \wedge c + (a \wedge b) \cdot c + a \wedge b \wedge c$$

$$(3.13)$$

$$a(bc) = a(b \cdot c + b \wedge c)$$
$$= a(b \cdot c) + a(b \wedge c)$$
$$= a \cdot b \cdot c + a \wedge (b \cdot c) + a \cdot (b \wedge c) + a \wedge b \wedge c$$

□

Note 3.2. $(a \cdot b) \wedge c + (a \wedge b) \cdot c = abc - cba$ and $a \wedge (b \cdot c) + a \cdot (b \wedge c) = abc - cba$.

Example 3.12. Let three elements $i, j,$ and $\alpha k$, where $i, j, k \in \mathbb{G}_3$ and $\alpha \in \mathbb{R}$ [1]. (i) Determine $i(j\alpha k)$. (ii) Determine $(ij)\alpha k$. (iii) Is the associativity fulfilled?

**Solution 3.12.** (i) $i(j\alpha k) = \sigma_1(\sigma_2 \alpha \sigma_3) = \alpha \sigma_{123}$. (ii) $(ij)\alpha k = (\sigma_1 \sigma_2)\alpha \sigma_3 = \alpha \sigma_{123}$. (iii) From the results of (i) and (ii) yes, it is.

## 3.2.5.  Reversion: $a^\dagger$

**Definition 3.8.** For an element $a \in \mathbb{G}_3$ [1], we define the reversion of the element $a$ as

$$a^\dagger$$

where $a^\dagger = (a_1 a_2 a_3 \cdots a_r)^\dagger = a_r \cdots a_3 a_2 a_1$.

Example 3.13. Let an element $a = 1 + 2\sigma_2 - 3\sigma_{12} + \sigma_3\sigma_2\sigma_1, \in \mathbb{G}_3$. Obtain the reverse of element $a$.

**Solution 3.13.** From the definition, the reversion of $a$ is $a^\dagger = 1 + 2\sigma_2 - 3\sigma_{21} + \sigma_1\sigma_2\sigma_3$.

## 3.2.6.  Dual: $Ia_r$

**Definition 3.9.** Let two elements $a = a_0 + a_1\sigma_1 + a_2\sigma_2 + a_3\sigma_3 + a_4\sigma_{12} + a_5\sigma_{23} + a_6\sigma_{13} + a_7\sigma_{123}$ and $b = b_0 + b_1\sigma_1 + b_2\sigma_2 + b_3\sigma_3 + a_4\sigma_{12} + a_5\sigma_{23} + a_6\sigma_{13} + a_7\sigma_{123}$ on $\mathbb{G}_3$, then

$$Ia_r = b_{n-r}$$

with $n = 3$

$$
\begin{array}{llll}
\text{scalar} & r = 0, & n-r = 3 & \text{trivector} \\
\text{vector} & r = 1, & n-r = 2 & \text{bivector} \\
\text{bivector } & r = 2, & n-r = 1 & \text{vector} \\
\text{trivector } & r = 3, & n-r = 0 & \text{scalar}
\end{array}
$$

$$(3.14)$$

From Eq. 3.14

$$\text{scalar} \rightleftarrows \text{trivector}$$

$$\text{vector} \rightleftarrows \text{bivector}$$

**Example 3.14.** Explain the possibilities of the duals on $\mathbb{G}_3$ [17, 20, 23].

**Solution 3.14.** We have four possibilities: (i) If $r = 0$ and $n = 3$, then $Ia_0 = b_3$. (ii) If $r = 1$ and $n = 3$, then $Ia_1 = b_2$. (iii) If $r = 2$ and $n = 3$, then $Ia_2 = b_1$. And (iv) If $r = 3$ and $n = 3$, then $Ia_3 = b_0$. From these possibilities, the only duals are scalar $\rightleftarrows$ trivector and vector $\rightleftarrows$ bivector.

**Example 3.15.** Let two elements $a = \sigma_1 + \sigma_2$ and $b = 2\sigma_1 + \sigma_2 \in \mathbb{G}_3$. Obtain the dual of the bivector $a \wedge b$.

**Solution 3.15.** $a \wedge b = -\sigma_{12}$, then $I(a \wedge b) = \sigma_{123}(-\sigma_{12}) = -\sigma_{23}\sigma_2 = \sigma_3$.

## 3.2.7.  Blades: $< a >$

**Definition 3.10.** For an element $a = a_0 + a_1\sigma_1 + a_2\sigma_2 + a_3\sigma_3 + a_4\sigma_{12} + a_5\sigma_{23} + a_6\sigma_{13} + a_7\sigma_{123}$ on $\mathbb{G}_3$, a vector space structure is determined using Def. 3.15 [17, 20, 23].

$$a = \underbrace{a_0}_{grade-0\,blade} + \underbrace{a_1\sigma_1 + a_2\sigma_2 + a_3\sigma_3}_{grade-1\,blades} + \underbrace{a_4\sigma_{12} + a_5\sigma_{23} + a_6\sigma_{13}}_{grade-2\,blades} + \underbrace{a_7\sigma_{123}}_{grade-3\,blade}$$

$$(3.15)$$

where element $a$ can be expressed as

$$a = \sum_{r=1}^{3} < a >_i \;\; = \;\; < a >_0 + < a >_1 + < a >_2 + < a >_3$$

**Corollary 3.1.** Any element $a$ can be separated into the sum of even and odd blades.

$$< a > = < a >_+ + < a >_-$$

$$(3.16)$$

$$< a >_+ = < a >_0 + < a >_2 + \cdots +$$
$$< a >_- = < a >_1 + < a >_3 + \cdots +$$

**Example 3.16.** Let an element $a = 1 + 2\sigma_1 - 3\sigma_2 - 3\sigma_2\sigma_1, \in \mathbb{G}_3$. Obtain the blades of element $a$.

**Solution 3.16.** From the definition, $< a >_0 = 1$, $< a >_1 = 2\sigma_1 - 3\sigma_2$, and $< a >_2 = -3\sigma_2\sigma_1$.

## 3.2.8.   Norm: $||a||$

**Definition 3.11.** Let the element $a = a_0 + a_1\sigma_1 + a_2\sigma_2 + a_3\sigma_3 + a_4\sigma_{12} + a_5\sigma_{23} + a_6\sigma_{13} + a_7\sigma_{123}$ on $\mathbb{G}_3$. Its **norm** is the scalar part, in symbols $\lfloor x \rfloor$), of Eq. 3.17 [17, 20, 23].

$$||a|| = \sqrt{\lfloor aa^\dagger \rfloor} \tag{3.17}$$

Example 3.17. Let an element $a = -1 + 2\sigma_1 + 3\sigma_{123} \in \mathbb{G}_3$. Obtain the norm of element $a$.

**Solution 3.17.** From the definition, the norm of $a$ is Eq. 3.18

$$
\begin{aligned}
||a|| &= \sqrt{\lfloor aa^\dagger \rfloor} \\
&= \sqrt{\lfloor (-1 + 2\sigma_1 + 3\sigma_{123})(-1 + 2\sigma_1 - 3\sigma_{123}) \rfloor} \\
&= \sqrt{\lfloor 14 + \cancel{3\sigma_1} \rfloor} \\
&= \sqrt{14}.
\end{aligned}
\tag{3.18}
$$

## 3.2.9.   Vector Components: $v_{\parallel} \ v_{\perp}$

The components of a vector with the form $a = a_1\sigma_1 + a_2\sigma_2 + a_3\sigma_3$ on $\mathbb{G}_3$, $v = v_1\sigma_1 + v_2\sigma_2 + v_3\sigma_3$ on $\mathbb{G}_3$, are the orthonormal projection of vector $v$ onto vector $a$ $v_{\parallel}$ and the orthonormal projection of vector $v$ onto vector $a$ $v_{\perp}$ (Fig. 3.3), such that vector $v = v_{\parallel} + v_{\perp}$ [25].

$$v_{\parallel} = \frac{v \cdot a}{||a||^2} a \tag{3.19}$$

$$v_{\perp} = v - v_{\parallel} \tag{3.20}$$

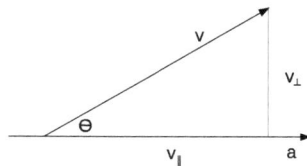

**Figure 3.3** Components of vector $b$.

Example 3.18. Let the vectors $a = 2\sigma_2$ and $v = \sigma_1 + 4\sigma_3$ on $G_2$. Determine $v_{\parallel}$ and $v_{\perp}$.

**Solution 3.18.** $v \cdot a = 2\sigma_{12} - 8\sigma_{23}$, then $v_{\parallel} = \dfrac{v \cdot a}{||a||^2} a = \dfrac{2}{2}(\sigma_{12} - 4\sigma_{23}) = \sigma_{12} - 4\sigma_{23}$. $v_{\perp} = v - v_{\parallel} = \sigma_1 + 4\sigma_3 - \sigma_{12} + 4\sigma_{23}$.

## 3.3.  Reflections and Rotations in Space

Given a **vector** $a = a_1\sigma_1 + a_2\sigma_2 + a_3\sigma_3 \in G_3$, we define a **reflection** of vector $a$ as $Ia$ or $IIa$, where $I = \sigma_1 \wedge \sigma_2 \wedge \sigma_3$ [1, 2, 9, 20, 22, 23, 24, 26]. Here, a **rotation** is defined as two successive reflections.

Example 3.19. Let an element $a = \sigma_1 + \sigma_2 + \sigma_3 \in G_3$ [1]. (i) Determine $Ia$. (ii) Determine $aI$. (iii) Explain (i) and (ii).

**Solution 3.19.** (i) $Ia = \sigma_1\sigma_2\sigma_3 a = \sigma_{23} - \sigma_{13} + \sigma_{12}$. (ii) $aI = a\sigma_1\sigma_2\sigma_3 = \sigma_{23} - \sigma_{13} + \sigma_1 2$. (iii) From these results, (i) is a **rotation** of $\frac{\pi}{2}$ in the **clockwise** direction and (ii) is a **rotation** of $\frac{\pi}{2}$ in the **counter-clockwise** direction.

Example 3.20. Let an element $a = \sigma_1 + \sigma_2 \in G_3$ [1]. (i) Determine $IIa$. (ii) Determine $aII$. (iii) Explain (i) and (ii).

**Solution 3.20.** (i) $IIa = \sigma_1\sigma_2\sigma_1\sigma_2 a = -\sigma_1 - \sigma_2$. (ii) $aII = a\sigma_1\sigma_2\sigma_1\sigma_2 = -\sigma_1 - \sigma_2$. (iii) From these results, (i) is a **reflection** of $\pi$ in the **clockwise** direction and (ii) is a **reflection** of $\pi$ in the **counter-clockwise** direction.

An alternative procedure is to find vector $y = -uxu$, where $u = \dfrac{u}{||u||}$ is a reflection vector and vector $x$ is the vector to be transformed.

To rotate an angle $\theta$ in a plane, perform two successive reflections along the two axes in the plane that satisfy these conditions: (a) that the angle between the two axes is $\dfrac{\theta}{2}$ and (b) that the rotation from the first axis to the second is in the same direction as the rotation to be performed (Figs. 2.5-2.7).

Thus, to rotate vector $v$ let $I = \sigma_{12}$ and let $m$ and $n$ be vectors along the axes satisfying the conditions, so the result of the rotation is

$$
\begin{aligned}
v' &= -m(-nvn^{-1})m^{-1} \\
&= (mn)v(mn)^{-1} \\
&= uvu^{-1}
\end{aligned}
\tag{3.21}
$$

where $u = mn$

Example 3.21. Find the vector $y$ for vector $x = \sigma_1 + \sigma_2 + 3\sigma_3$ to $u = \sigma_1 + \sigma_2 + \sigma_3$.

**Solution 3.21.** $u = \dfrac{u}{||u||} = \dfrac{u}{2}$, then $y = -uxu = -\dfrac{3}{2}\sigma_1 - \dfrac{1}{2}\sigma_2$.

## 3.4.   Geometric Representation of a Line on $\mathbb{R}^3$

**Definition 3.12.** Given **vector** $v$ and a point $x_0$ on $\mathbb{R}^3$, what is the equation of the line passing through the point $x_0$ in the direction of vector $v$ (Fig. 3.4)? [2, 9, 20, 22, 24] (Example adapted with permission of the author [22], and reproduced here from [1]).

The line $L_{x_0}(v)$ is given by

$$L_{x_0}(v) := \{\mathbf{x} \,|\, (\mathbf{x} - x_0) \wedge v = 0\} \tag{3.22}$$

Remark 3.8. Note that the oriented line is defined by $v$.

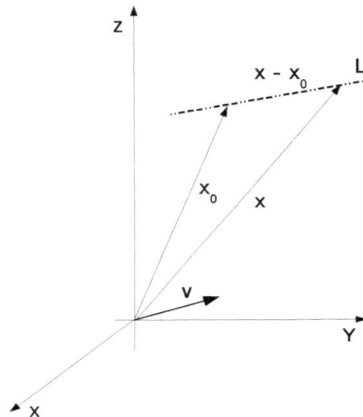

**Figure 3.4** Line $L_{x_0}$ through the point $x_0$ in the direction of $v$. (Figure taken with the permission of the author) [1])

Example 3.22. Given vector $v$ and a point $x_0$ in the space $\mathbb{R}^3$, what is the equation of the line passing through the point $x_0 = (1, 2, 3)$ in the direction of the vector $v = (1, 1, 1)$?

**Solution 3.22.** The line $L_{x_0}(v)$ is given by

$$L_{x_0 = (1,2,3)}(v) := \{\mathbf{x} \,|\, (\mathbf{x} - x_0) \wedge v = 0\} \tag{3.23}$$

$$
\begin{aligned}
&[(x_1 \sigma_1 + x_2 \sigma_2 + x_3 \sigma_3) - (\sigma_1 + 2\sigma_2 + 3\sigma_3)] \wedge (\sigma_1 + \sigma_2 + \sigma_3) = 0 \\
&[(x_1 - 1)\sigma_1 + (x_2 - 2)\sigma_2 + (x_3 - 3)\sigma_3] \wedge (\sigma_1 + \sigma_2 + \sigma_3) = 0
\end{aligned} \tag{3.24}
$$

The outer product $(x - x_0) \wedge v = \frac{1}{2}[(x - x_0)v - v(x - x_0)]$,

$$[(x_1-1)\sigma_1+(x_2-2)\sigma_2+(x_3-3)\sigma_3](\sigma_1+\sigma_2+\sigma_3) = (x_1+x_2+x_3-6)$$
$$+ (x_1-x_2+1)\sigma_{12}$$
$$+ (x_1-x_3+2)\sigma_{13}$$
$$+ (x_2-x_3+1)\sigma_{23}$$

$$(\sigma_1+\sigma_2+\sigma_3)[(x_1-1)\sigma_1+(x_2-2)\sigma_2+(x_3-3)\sigma_3] = (x_1+x_2+x_3-6)$$
$$+ (x_2-x_1-1)\sigma_{12}$$
$$+ (x_3-x_1-2)\sigma_{13}$$
$$+ (x_3-x_2-1)\sigma_{23}$$

$$(3.25)$$

$$x_1 - x_2 + 1 = 0 \qquad\qquad (3.26)$$
$$x_1 - x_3 + 2 = 0$$
$$x_2 - x_3 + 1 = 0$$

From Eqs. 4.18, the system has multiple solutions $x_1 - x_3 = -2$ and $x_2 - x_3 = -1$; so, the points with the form $(x_1, x_3 - 1, x_1 + 2)$ are the solution. Note that the point $(1, 2, 3)$ meets the line $L_{x_0}(v)$.

## 3.5.   Geometric Representation of a Plane on $\mathbb{R}^3$

**Definition 3.13.** Given two **vectors** $v$, $u$, and a point $x_0$ in $\mathbb{R}^3$, what is the equation of the plane passing through the point $x_0$ over the plane generated by the vectors $v$ and $u$? (Fig. 3.5) (Case adapted with permission of the author [1, 22]).
The plane $P_{x_0}(u \wedge v)$ is given by

$$P_{x_0}(u \wedge v) := \{\mathbf{x} \,|\, (\mathbf{x} - x_0) \wedge (u \wedge v) = 0\} \qquad (3.27)$$

Remark 3.9. Note that the oriented plane is defined by $u \wedge v$.

Plane := $(x-x_0) \wedge (u \wedge v) = 0$

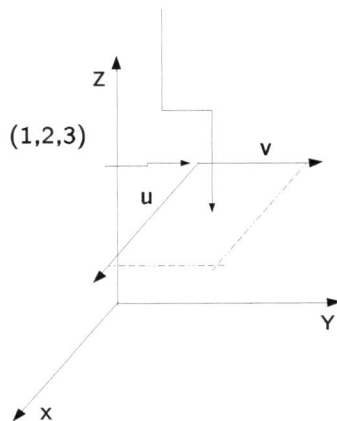

**Figure 3.5** The plane $P_{x_0}$ through the point $x_0$ in the direction of the bivector $u \wedge v$.

Example 3.23. Given two vectors $v$, $u$, and a point $x_0$ in the space $\mathbb{R}^3$, what is the equation of the plane passing through the point $x_0 = (1,2,3)$ in the plane generated by the vectors $v = (1,0,0)$ and $u = (0,1,0)$?

**Solution 3.23.** The plane $P_{x_0}(u,v)$ is given by

$$P_{x_0=(1,2,3)}(u \wedge v) := \{\mathbf{x} \mid (\mathbf{x} - x_0) \wedge (u \wedge v) = 0\} \tag{3.28}$$

$$\begin{aligned}
[(x_1\sigma_1 + x_2\sigma_2 + x_3\sigma_3) - (\sigma_1 + 2\sigma_2 + 3\sigma_3)] \wedge (\sigma_1\sigma_2) = 0 \\
[(x_1 - 1)\sigma_1 + (x_2 - 2)\sigma_2 + (x_3 - 3)\sigma_3] \wedge (\sigma_1\sigma_2) = 0
\end{aligned} \tag{3.29}$$

From Eq. 3.29, $[(x_1 - 1)\sigma_1 + (x_2 - 2)\sigma_2 + (x_3 - 3)\sigma_3][\sigma_1\sigma_2] - [\sigma_1\sigma_2][(x_1 - 1)\sigma_1 + (x_2 - 2)\sigma_2 + (x_3 - 3)\sigma_3] = 0$; so, the points with the form $(2 - x_2, 1 - x_1, 3)$ are the solution. Note that the point $(1,2,3)$ meets the plane $P_{x_0=(1,2,3)}(u \wedge v)$.

## 3.6. Remarks

Note that the outer product $a \wedge b \wedge c$ is neither a scalar nor a vector. In fact, it is a region in the space $\mathbb{R}^3$, spanned by the vectors $a$, $b$, and $c$ that represent a region with a volume and an orientation; (interchange $a$, $b$, and $c$ and you will change the sign of $a \wedge b \wedge c$) [2, 9, 20, 22, 23, 24, 26]. This operator makes unnecessary the **cross product** that is usually used in the **Heaviside-Gibbs algebra** to identify the areas of a region in the space $\mathbb{R}^3$, for this reason, in Chap. 1 we only used the **cross product**. In this chapter, we showed the concordance between both algebras in the space $\mathbb{R}^3$. In the next chapter we will elaborate on **Geometric algebra** or **Grassmann algebra** in the space $\mathbb{R}^n$.

The **Geometric algebra** or **Grassmann algebra** acts in an $n$ dimensional space, which enables the extension of all operators to that dimension.

It is important that the reader reviews all the exercises in this chapter and solves them, to get familiar with this algebra.

## 3.7. Exercises

**Exercise 3.1.** Provide some exercises of elements on $\mathbb{G}_3$.

**Exercise 3.2.** Let two elements $a = (-1, 1, 1)$ and $b = (2, 3, -3) \in \mathbb{G}_3$ [1]. Express these elements using the orthonormal basis.

**Exercise 3.3.** Consider two elements $a = \sigma_{123}$ and $b = \sigma_1 + \sigma_2 \in \mathbb{G}_3$. Obtain the outer product $a \wedge b$.

**Exercise 3.4.** Let the vectors $a = -\sigma_1 + 2\sigma_3$ and $b = \sigma_1 + \sigma_2 \in \mathbb{G}_3$. Determine the outer product.

**Exercise 3.5.** Consider elements $a = \sigma_3, b = -\sigma_{213} \in \mathbb{G}_3$. (i) Obtain the geometric products $ab$ and $ba$. (ii) Determine $a \cdot b$. (iii) Determine $a \wedge b$.

**Exercise 3.6.** Let three elements $a = \sigma_3, b = \sigma_1 - \sigma_2$, and $c = \sigma_1 + \sigma_{123} \in \mathbb{G}_3$ [1]. (i) Determine $a(b+c)$. (ii) Determine $ab + ac$. (iii) Is the distributivity fulfilled?

**Exercise 3.7.** Let three elements $a = \sigma_{123}, b = 3 + \sigma_1 - \sigma_2$, and $c = -\sigma_{12} \in \mathbb{G}_3$. (i) Determine $a(b+c)$ and $(b+c)a$. (ii) Determine $a \wedge (b+c)$. (iii) Determine $a \wedge b$ and $a \wedge c$. (iv) Is the distributivity over the outer product fulfilled?

**Exercise 3.8.** Let an element $a = 1 - \sigma_1 + 2\sigma_1\sigma_3 \in \mathbb{G}_3$. Obtain the element $a^{-1}$.

**Exercise 3.9.** Let an element $a = \sigma_1\sigma_3 \in \mathbb{G}_3$. Obtain the reverse of element $a$.

**Exercise 3.10.** Let an element $a = 1 + 2\sigma_1\sigma_2 - \sigma_{123} \in \mathbb{G}_3$. Obtain the blades.

**Exercise 3.11.** Let two elements $a = \sigma_1\sigma_3$ and $b = 2\sigma_1 + \sigma_2 \in \mathbb{G}_3$. Obtain the dual of $a \wedge b$.

**Exercise 3.12.** Let an element $a = 1 + \sigma_1 + \sigma_2 + \sigma_2\sigma_1 \in \mathbb{G}_3$. Obtain the norm of element $a$.

**Exercise 3.13.** Consider three elements $a = \sigma_3, b = \sigma_{12}$, and $c = \alpha\sigma_1 - \sigma_2 \in \mathbb{G}_3$, where $\alpha \in \mathbb{R}$. (i) Determine $a(bc)$. (ii) Determine $(ab)c$. (iii) Is the associativity fulfilled?

**Exercise 3.14.** Let an element $a = \sigma_1 + \sigma_3 \in \mathbb{G}_3$ [1]. (i) Determine $Ia$. (ii) Determine $aI$. (iii) Explain (i) and (ii).

**Exercise 3.15.** Let an element $a = \sigma_2 + \sigma_3 \in \mathbb{G}_3$ [1]. (i) Determine $IIa$. (ii) Determine $aII$. (iii) Explain (i) and (ii).

**Exercise 3.16.** Find vector $y$ for the vector $x = 2\sigma_1 + \sigma_3$ onto vector $u = \sigma_1 - 2\sigma_2$.

**Exercise 3.17.** Given a vector $v$ and a point $x_0$ on the plane $\mathbb{R}^3$. What is the equation of the **line** passing through the point $x_0 = (0, 1, 0)$ in the direction of the vector $v = (1, 1, 1)$?

**Exercise 3.18.** Given two vectors $v$ and $u$, and a point $x_0$ in the space $\mathbb{R}^2$. What is the equation of the **plane** passing through point $x_0 = (2, 1, 1)$ in the plane generated by the vectors $v = (0, 1, 1)$ and $u = (1, 1, 1)$?

# CHAPTER 4

# Geometric Algebra on $\mathbb{G}_n$

**Carlos Polanco**

Faculty of Sciences, Universidad Nacional Autónoma de México, México

**Abstract** This chapter reviews and elaborates on the operators of Geometric algebra from $\mathbb{G}_3$ to $\mathbb{G}_n$. This algebra is attributed to Hermann Grassmann [Die lineare Ausdehnungslehre, ein neuer Zweig der Mathematik 1842]. It is formed by two main operators, the **outer product** and **inner product**. Here, a new element is introduced the **multivector**, we review these operators, their properties, and their application in the representation of curves, planes, and objects on space $\mathbb{G}_n$.

**Keywords**: Associativity: $a(bc) = (ab)c$, bivector: $a \wedge b$, blades $< a >$, component: $v_{\parallel}$, component: $v_{\perp}$, distributivity: $a(b+c)$, distributivity: $a \wedge (b+c)$, dual $Ia_r = b_{n-r}$, equation of a line, outer product, geometric algebra, geometric product, inner product, lines, multiplicative inverse: $a^{-1}$, multivector $a \wedge b \wedge c \wedge \cdots \wedge z$, norm $||a||$, reflections, reversion: $a^{\dagger}$, rotations, trivector: $a \wedge b \wedge c$

## 4.1.   Preliminaries

This chapter explores the main operators in space $\mathbb{G}_n$, since this space corresponds to $\mathbb{G}_n = \mathbb{G}_n(\mathbb{R}^n)$, we will **not** provide illustrative graphs, but we will focus on the analytical solutions oriented to the elements in that space using the **outer product** $a \wedge b$.

Note 4.1. It is important to note that although the elements $\mathbb{G}_n$ have $\sigma_{1 \cdots n}$ (Def. 4.1), to simplify, we have replaced them with examples on $\mathbb{G}_4$.

## 4.2.   Geometric Algebra on $\mathbb{G}_n$

**Definition 4.1.** The **Geometric algebra** or **Grassmann algebra** [1, 9] is a unitary associative algebra, in symbols $\mathbb{G}_n = \mathbb{G}_n(\mathbb{R}^n)$. It is formed by $2^n$ elements:

**scalars** $\alpha_i$, **vectors** $\alpha_i \sigma_i$, **bivectors** $\alpha_{ij} \sigma_{ij}$, **trivectors** $\alpha_{ijk} \sigma_{ijk}$, and **multivectors** $\alpha_{i\cdots n} \sigma_{i\cdots n}$. For convenience, these elements are expressed in **orthonormal** basis that meet Eqs. 4.1 for $i \neq j$.

$$\sigma_i \sigma_i = 1$$
$$\sigma_i \sigma_j = -\sigma_j \sigma_i$$

(4.1)

An arbitrary element is (Eq. 4.2).

$$v = \underbrace{\sum_{i=1}^{n} v_i}_{\text{basis scalars}} + \underbrace{\sum_{i=1}^{n} v_i \sigma_i}_{\text{basis vectors}} + \underbrace{\sum_{i,j=1}^{n} v_{ij} \sigma_{ij}}_{\text{basis bivectors}}$$
$$+ \underbrace{\sum_{i,j,k=1}^{n} v_{ijk} \sigma_{ijk}}_{\text{basis trivectors}} + \underbrace{\sum_{i,\cdots,z=1}^{n} v_{i\cdots z} \sigma_{i\cdots z}}_{\text{basis multivector}} \text{ in } \mathbb{G}_n.$$

(4.2)

Remark 4.1. The equivalent is $\sigma_i \wedge \sigma_j$, $\sigma_i \sigma_j$ and $\sigma_{ij}$.

Example 4.1. Provide some examples of elements on $\mathbb{G}_n$.

**Solution 4.1.** $v = 5\sigma_{1\cdots n}$,

# 4.2.1.　Outer Product: $a \wedge b$

**Definition 4.2.** For two elements $a$ and $b \in \mathbb{G}_n$ [1], we define

$$a \wedge b = \frac{1}{2}(ab - ba)$$

Remark 4.2. If $a \wedge b = 0 \Rightarrow a \parallel b$.

Example 4.2. Consider two elements $a = \sigma_{1234}$ and $b = 2 + \sigma_{12} \in \mathbb{G}_n$. Obtain the outer product $a \wedge b$.

**Solution 4.2.** $ab = 2\sigma_{1234} + \sigma_{123412} = 2\sigma_{1234} - \sigma_{34}$ and $ba = 2\sigma_{1234} - \sigma_{34}$, so $a \wedge b = 2\sigma_{34}$.

Example 4.3. Let two elements $a$ and $b$ on $\mathbb{G}_n$ $a = \sigma_5 + \sigma_1$, where $b = \alpha \sigma_5 + \beta \sigma_1$. Determine what values comply with the scalars $\alpha$ and $\beta$ so both vectors are collinear.

**Solution 4.3.** If $a \wedge b = \frac{1}{2}(ab - ba) = 0$, then $a \parallel b$. Since $ab = 2\alpha$, $ba = 2\beta$

$a \wedge b = \frac{1}{2}(ab - ba) = 0 \Leftrightarrow 2\alpha - 2\beta = 0 \Leftrightarrow \alpha = \beta$.

## 4.2.2.   Inner Product: $a \cdot b$

**Definition 4.3.** For two elements $a$ and $b \in \mathbb{G}_n$ [1], we define

$$a \cdot b = \frac{1}{2}(ab + ba).$$

Remark 4.3. If $a \cdot b = 0 \Leftrightarrow a \perp b$.

Example 4.4. Let two elements $a = \sigma_{567}, b = \sigma_{12345} \in \mathbb{G}_n$. (i) Obtain the geometric products $ab$ and $ba$. (ii) Determine $a \cdot b$. (iii) Determine $a \wedge b$.

**Solution 4.4.** (i) $ab = -\sigma_{123467}, ba = \sigma_{123467}$. (ii) $a \cdot b = 0$. (iii) $a \wedge b = -\sigma_{123467}$.

Example 4.5. Let two elements $a = \sigma_1 \sigma_2$ and $b = \sigma_3 \in \mathbb{G}_n$. (i) Obtain the geometric products $ab$ and $ba$. (ii) From Def. 4.3, determine $a \cdot b$. (iii) From Def. 4.2, determine $a \wedge b$.

**Solution 4.5.** (i) $ab = \sigma_1 \sigma_2 \sigma_3$. $ba = \sigma_1 \sigma_2 \sigma_3$. (ii) $a \cdot b = \sigma_1 \sigma_2 \sigma_3$. (iii) $a \wedge b = 0$.

Example 4.6. Let two elements $a = 1 + \sigma_1 + \sigma_2 - \sigma_{24}$ and $b = \sigma_{1234} \in \mathbb{G}_n$. (i) Obtain the geometric products $ab$ and $ba$. (ii) From Def. 4.3, determine $a \cdot b$. (iii) From Def. 4.2, determine $a \wedge b$.

**Solution 4.6.** (i) $ab = (\sigma_{1234})(\sigma_1 + \sigma_2 - \sigma_{24}) = \sigma_{234} - \sigma_{134} - \sigma_{13}$. $ba = -\sigma_{234} + \sigma_{134} - \sigma_{13}$ (ii) $a \cdot b = -\sigma_{13}$. (iii) $a \wedge b = \sigma_{234} - \sigma_{134}$.

Example 4.7. Let two elements $a$ and $b$ on $\mathbb{G}_n$ [1] $a = \sigma_1 + \sigma_2 + \sigma_3 + \sigma_4$ and $b = \sigma_1 + \sigma_2 - 2\sigma_3 + \sigma_4$ on $\mathbb{G}_n$. (i) Determine the inner product. (ii) Geometrically interpret the **inner product**.

**Solution 4.7.** (i) $ab = 1 - 2\sigma_{13} + \sigma_{23} + 3\sigma_{34}$ and $ba = 1 + 3\sigma_{13} + 2\sigma_{23} - 3\sigma_{34}$, $a \cdot b = \frac{1}{2}(ab + ba) = 1$. (ii) It is a point in the $\mathbb{G}_4$ space.

## 4.2.3.   Geometric Product: $ab$

From two elements $a$ and $b \in \mathbb{G}_n$, the **geometric product** (Eq. 4.3) is defined as [1]

$$\mathbf{ab} = a \cdot b + a \wedge b \tag{4.3}$$

where the term $\mathbf{a} \cdot \mathbf{b}$ is the **inner product** (Def. 4.2.2) and the term $\mathbf{a} \wedge \mathbf{b}$ is the **outer product** (Def. 4.2.1) [1].

Remark 4.4. If the elements on $\mathbb{G}_n$ are of the form $a = a_1\sigma_1 + a_2\sigma_2 + \cdots + a_n\sigma_3$ and $b = b_1\sigma_1 + b_2\sigma_2 + \cdots + b_n\sigma_n$, then the **inner product** will only have the **scalar part** and the **outer product** the **vectorial part**.

Note 4.2. The geometrical description of the **multivector** $\sigma_1 \wedge \sigma_2 \wedge \sigma_3 \wedge \cdots \wedge \sigma_n$ is the **oriented multi-volume** spanned by n-vectors in space $\mathbb{G}_n$.

Example 4.8. Let two elements $a$ and $b \in \mathbb{G}_n$ [1]. Discuss the results obtained in Eqs. 2.4-2.8 and Eqs. 3.4-3.8 in space $\mathbb{G}_2$.

**Solution 4.8.** The **geometric product** $ab$ is the addition of the **inner product** $a \cdot b$ and the **outer product** $a \wedge b$. But if the elements $a$ and $b$ are only formed by **basis vectors**, then the **inner product** is the **real part** of the geometric product $ab$ and the **vector part** corresponds to the **outer product**.

## 4.3.   Properties on $\mathbb{G}_n$

The following fundamental properties of Geometric algebra [20, 23] are exemplified, so it is recommended to review the examples provided.

### 4.3.1.   Distributivity: $a(b+c)$

**Definition 4.4.** For three elements $a$, $b$, and $c \in \mathbb{G}_n$ [23, 24], then

$$a(b+c) = ab + ac$$

Proof.

$$
\begin{aligned}
ab + ac &= a \cdot b + a \wedge b + a \cdot c + a \wedge c \\
&= a \cdot b + a \cdot c + a \wedge b + a \wedge c \\
&= a \cdot (b+c) + a \wedge (b+c) \\
&= a(b+c)
\end{aligned}
\tag{4.4}
$$

$\square$

Example 4.9. Let three elements $a = \sigma_{2345}, b = \sigma_1 - \sigma_2$, and $c = \sigma_1 + \sigma_2$. (i) Determine $a(b+c)$. (ii) Determine $ab + ac$. (iii) Is the distributivity fulfilled? (iv) Explain the final product.

**Solution 4.9.** (i) $a(b+c) = \sigma_{2345}(\sigma_1 - \sigma_2 + \sigma_1 + \sigma_2) = 2\sigma_{12345}$. (ii) $ab = \sigma_{12345} + \sigma_{345}$ and $ac = \sigma_{12345} - \sigma_{345}$, then $ab + ac = 2\sigma_{12345}$. (iii) From (i) and (ii) yes, it is. (iv) The product $a(b+c) = 2\sigma_{12345}$ is an object in space $G_5(\mathbb{R}^5)$.

### 4.3.2.   Distributivity: $a \wedge (b+c)$

**Definition 4.5.** For three elements $a$, $b$, and $c \in \mathbb{G}_n$ [1], then

$$a \wedge (b+c) = a \wedge b + a \wedge c$$

Proof.

$$a \wedge (b+c) = -ab - ac + a \cdot (b+c)$$
$$= -ab - ac + a \cdot b + a \cdot c$$
$$= (-ab + a \cdot b) + (-ac + a \cdot c) \tag{4.5}$$
$$= a \wedge b + a \wedge c$$

$\square$

**Example 4.10.** Let three elements $a = 1 + \sigma_{1234}, b = \sigma_1 \sigma_3$, and $c = \sigma_5$. (i) Determine $a \wedge (b+c)$. (ii) Determine $a \wedge b + a \wedge c$. (iii) Is the distributivity fulfilled?

**Solution 4.10.** (i) $a \wedge (b+c) = (1 + \sigma_{1234}) \wedge (\sigma_1 + \sigma_{13}) = 0$. (ii) $a \wedge b + a \wedge c = 0 + 0 = 0$. (iii) From (i) and (ii) yes, it is.

### 4.3.3. Multiplicative Inverse: $a^{-1}$

**Definition 4.6.** For an element $a \in \mathbb{G}_n$ [1], we define $a^{-1} = \dfrac{a}{a \cdot a}$
then

$$aa^{-1} = 1$$

Proof.

$$aa^{-1} = a \frac{a}{a \cdot a}$$

$$= \frac{aa}{a \cdot a}$$
$$= \frac{a \cdot a + a \wedge a}{a \cdot a} \tag{4.6}$$
$$= \frac{a \cdot a}{a \cdot a} + \frac{a \wedge a}{a \cdot a}$$
$$= 1 + 0$$
$$= 1$$

$\square$

**Example 4.11.** Let an element $a = 1 + 2\sigma_{14567} \in \mathbb{G}_n$. Obtain the elements $a^{-1}$ and $aa^{-1}$ [1].

**Solution 4.11.** $a^{-1} = \dfrac{a}{a \cdot a} = \dfrac{1 + 2\sigma_{14567}}{5 + 4\sigma_{14567}} \Rightarrow aa^{-1} = \dfrac{(1 + 2\sigma_{14567})^2}{5 + 4\sigma_{14567}}$
$= 1.$

### 4.3.4. Associativity: $a(bc) = (ab)c$

**Definition 4.7.** For three elements $a$, $b$, and $c \in \mathbb{G}_n$ [1], then

$$(ab)c = a(bc)$$

Proof.

$$(ab)c = (a \cdot b + a \wedge b)c$$
$$= (a \cdot b)c + (a \wedge b)c$$
$$= a \cdot b \cdot c + (a \cdot b) \wedge c + (a \wedge b) \cdot c + a \wedge b \wedge c$$

$$(4.7)$$

$$a(bc) = a(b \cdot c + b \wedge c)$$
$$= a(b \cdot c) + a(b \wedge c)$$
$$= a \cdot b \cdot c + a \wedge (b \cdot c) + a \cdot (b \wedge c) + a \wedge b \wedge c$$

$\square$

**Note 4.3.** $(a \cdot b) \wedge c + (a \wedge b) \cdot c = abc - cba$ and $a \wedge (b \cdot c) + a \cdot (b \wedge c) = abc - cba$

**Example 4.12.** Let three elements $i, j,$ and $\alpha j$, where $i, j \in \mathbb{G}_n$ and $\alpha \in \mathbb{R}$ [1]. (i) Determine $i(j\alpha j)$. (ii) Determine $(ij)\alpha j$. (iii) Is the associativity fulfilled?

**Solution 4.12.** (i) $i(j\alpha j) = \sigma_1(\sigma_2 \alpha \sigma_2) = \alpha \sigma_1$. (ii) $(ij)\alpha j = (\sigma_1 \sigma_2)\alpha \sigma_2 = \alpha \sigma_1$. (iii) From the results of (i) and (ii) yes, it is.

## 4.3.5.  Reversion: $a^\dagger$

**Definition 4.8.** For an element $a \in \mathbb{G}_2$ [1], we define the reversion of element $a$ as $a^\dagger$, where
$$a^\dagger = (a_1 \cdots a_r)^\dagger = a_r \cdots a_1$$

**Example 4.13.** Let an element $a = 1 + 2\sigma_2 - 3\sigma_{12} + \sigma_1 \sigma_2 \sigma_3 \sigma_4 \sigma_5, \in \mathbb{G}_n$. Obtain the reverse of element $a$.

**Solution 4.13.** From the definition, the reverse of $a$ is $a^\dagger = 1 + 2\sigma_2 - 3\sigma_{21} + \sigma_5 \sigma_4 \sigma_3 \sigma_2 \sigma_1$.

## 4.3.6.  Dual: $Ia_r$

**Definition 4.9.** Let two elements $a$ and $b$ on $\mathbb{G}_n$, then (Eq. 4.8).

$$Ia_r = b_{n-r} \qquad (4.8)$$

**Example 4.14.** Let two elements $a = \sigma_{12345}$ and $b = \sigma_1 \in \mathbb{G}_n$. Obtain the dual of the two elements.

**Solution 4.14.** The dual of $a$ is scalar $\rightleftarrows$ vector . The dual of $b$ are: scalar $\rightleftarrows$ 5-vector, vector $\rightleftarrows$ 4-vector, and bivector $\rightleftarrows$ trivector.

## 4.3.7. Blades: $< a >$

**Definition 4.10.** For an element $a$ and $b$ on $\mathbb{G}_n$, a graded vector space structure is established using Def. 4.9 [17, 20, 23]

$$a = \underbrace{a_0}_{grade-0\ blade} + \underbrace{a_1\sigma_1 + a_2\sigma_2 + a_3\sigma_3}_{grade-1\ blades} + \underbrace{a_4\sigma_{12} + a_5\sigma_{23} + a_6\sigma_{13}}_{grade-2\ blades} + \cdots + \underbrace{a_n\sigma_{1\cdots n}}_{grade-n\ blade} \tag{4.9}$$

where the element or **multivector** $a$ can be expressed as

$$a = \sum_{r=1}^{n} <a>_i \; = \; <a>_0 + <a>_1 + <a>_2 + \cdots + <a>_n$$

**Corollary 4.1.** Any element or **multivector** $a$ can be separated into the sum of even $(2n)$ and odd $(2n-1)$ blades, where $n \in \mathbb{Z}^+$.

$$< a > = < a >_+ + < a >_-$$

$$< a >_+ = < a >_0 + < a >_2 + \cdots + < a >_{2n} \tag{4.10}$$
$$< a >_- = < a >_1 + < a >_3 + \cdots + < a >_{2n-1}$$

Example 4.15. Let an element $a = 1 + 2\sigma_1 - 3\sigma_2 - 3\sigma_2\sigma_1\sigma_4\sigma_5, \in \mathbb{G}_n$. Obtain the blades of element $a$.

**Solution 4.15.** From the definition, $<a>_0 = 1$, $<a>_1 = 2\sigma_1 - 3\sigma_2$, and $<a>_4 = -3\sigma_2\sigma_1\sigma_4\sigma_5$.

## 4.3.8. Norm: $||a||$

**Definition 4.11.** Let the element $a$ on $\mathbb{G}_n$, its **norm** is the scalar part, in symbols $\lfloor x \rfloor$, of Eq. 4.11 [17, 20, 23].

$$||a|| = \sqrt{\lfloor aa^\dagger \rfloor} \tag{4.11}$$

Example 4.16. Let an element $a = 1 + 2\sigma_1 + 3\sigma_{12345} \in \mathbb{G}_5$. Obtain the norm of element $a$.

**Solution 4.16.** From the definition, the norm of $a$ is Eq. 4.12

$$\begin{aligned} ||a|| &= \sqrt{\lfloor aa^\dagger \rfloor} \\ &= \sqrt{\lfloor (1 + 2\sigma_1 + 3\sigma_{12345})(1 + 2\sigma_1 - 3\sigma_{12345}) \rfloor} \\ &= \sqrt{\lfloor 18 + \cancel{3\sigma_1} \rfloor} \\ &= \sqrt{18}. \end{aligned} \tag{4.12}$$

## 4.3.9.  Vector Components: $v_\parallel$ $v_\perp$

The components of a vector of the form $a = a_1\sigma_1 + a_2\sigma_2 + a_3\sigma_3 + \cdots + a_n\sigma_n$, $v = v_1\sigma_1 + v_2\sigma_2 + v_3\sigma_3 + \cdots + v_n\sigma_n$ on $\mathbb{G}_n$, are the orthonormal projection of vector $v$ onto vector $a$ $v_\parallel$ and the orthonormal projection of vector $v$ onto vector $a$ $v_\perp$, such that vector $v = v_\parallel + v_\perp$ [25].

$$v_\parallel = \frac{v \cdot a}{||a||^2} a \tag{4.13}$$

$$v_\perp = v - v_\parallel \tag{4.14}$$

Example 4.17. Let the vectors $v = \sigma_1 + 2\sigma_2 + \sigma_3 - \sigma_4$ and $a = \sigma_1 + 4\sigma_2 - \sigma_3 + 2\sigma_4$ on $G_4$. Determine $v_\parallel$ and $v_\perp$.

**Solution 4.17.** $v \cdot a = 6$ then $v_\parallel = \dfrac{v \cdot a}{||a||^2} a = \dfrac{3}{7}(\sigma_1 + 4\sigma_2 - \sigma_3 + 2\sigma_4)$. $v_\perp = v - v_\parallel$
$= \dfrac{4}{7}\sigma_1 + \dfrac{2}{7}\sigma_2 + \dfrac{4}{7}\sigma_3 - \dfrac{1}{7}\sigma_4$.

## 4.4.  Reflections and Rotations on $\mathbb{G}_n$

Given a **vector** $a \in \mathbb{G}_n$, we define a **reflection** of vector $a$ as $Ia$ or $IIa$, where $I = \sigma_1 \wedge \sigma_2 \wedge \sigma_3 \wedge \cdots \wedge \sigma_n$ [1, 2, 9, 20, 22, 23, 24, 26]. A **rotation** is here defined, as two successive reflections.

Example 4.18. Let an element $a = \sigma_1 + 2\sigma_2 + 3\sigma_3 + 4\sigma_4 + 5\sigma_5 \in \mathbb{G}_n$ [1]. (i) Determine $Ia$. (ii) Determine $aI$. (iii) Explain (i) and (ii).

**Solution 4.18.** (i) $Ia = \sigma_{2345} - 2\sigma_{1345} + 3\sigma_{1245} - 4\sigma_{1235} + 5\sigma_{1234}$. (ii) $aI = \sigma_{2345} - 2\sigma_{1345} + 3\sigma_{1245} - 4\sigma_{1235} + 5\sigma_{1234}$. (iii) From these results, (i) is the **rotation** of $\frac{\pi}{2}$ in the **clockwise** direction and (ii) is the **rotation** of $\frac{\pi}{2}$ in the **counter-clockwise** direction.

Example 4.19. Let an element $a = \sigma_1 + 2\sigma_2 + 3\sigma_3 + 4\sigma_4 + 5\sigma_5 \in \mathbb{G}_n$ [1]. (i) Determine $IIa$. (ii) Determine $aII$. (iii) Explain (i) and (ii).

**Solution 4.19.** (i) $IIa = a$. (ii) $aII = a$. (iii) From these results, (i) is the **reflection** of $2\pi$ in the **clockwise** direction and (ii) is the **reflection** of $2\pi$ in the **counter-clockwise** direction.

An alternative procedure is to find vector $y = -uxu$, where $u = \dfrac{u}{||u||}$ is the reflection vector and vector $x$ is the vector to be transformed. (Case taken with the permission of the author [2])

To rotate an angle $\theta$ in a plane, perform two reflections in succession along any of the two axes in the plane that meet these conditions: (a) that the angle between the axes is $\dfrac{\theta}{2}$ and (b) the rotation from the first axis to the second is in the same direction as the rotation to be performed (Figs. 2.5-2.7).

To rotate vector $v$, $I = \sigma_{12}$, let $m$ and $n$ be vectors along the axes satisfying the conditions, so the result of the rotation is

$$
\begin{aligned}
v' &= -m(-nvn^{-1})m^{-1} \\
&= (mn)v(mn)^{-1} \\
&= uvu^{-1}
\end{aligned}
\tag{4.15}
$$

where $u = mn$

**Example 4.20.** Find vector $y$ for the vector $x = \sigma_1 + 2\sigma_2 + 3\sigma_3 + 4\sigma_4 + 5\sigma_5$ onto vector $u = \sigma_1 + \sigma_2 + \sigma_3 + \sigma_4 + \sigma_5$.

**Solution 4.20.** $u = \dfrac{u}{||u||} = \dfrac{u}{5}$, so $y = -uxu = \dfrac{1}{5}(25\sigma_1 + 17\sigma_2 - 2\sigma_3 + 9\sigma_4 + 6\sigma_5 + 2\sigma_{123} + 2\sigma_{124} - 4\sigma_{125} - \sigma_{234} + \sigma_{235} + 2\sigma_{245} + 5\sigma_{134} + 6\sigma_{135} + 3\sigma_{345} + 2\sigma_{124} - 7\sigma_{145})$.

## 4.5. Analytical Representation of a Line in $\mathbb{R}^n$

**Definition 4.12.** Given a **vector** $v$ and a point $x_0$ in the $\mathbb{R}^n$ space, what is the equation of the line passing through the point $x_0$ in the direction of vector $v$? [2, 9, 20, 22, 24] (Example adapted with permission of the author [22], and reproduced here from [1]).
The line $L_{x_0}(v)$ is given by

$$
L_{x_0}(v) := \{\mathbf{x} \mid (\mathbf{x} - x_0) \wedge v = 0\}
\tag{4.16}
$$

Remark 4.5. Note that the line is defined by $v$.

**Example 4.21.** Given a vector $v$ and a point $x_0$ in space $\mathbb{R}^n$, what is the equation of the line passing through the point $x_0 = (1, 2, 3, 4)$ in the direction of the vector $v = (1, 1, 1, 1)$?

**Solution 4.21.** The line $L_{x_0}(v)$ is given by

$$
L_{x_0=(1,2,3,4)}(v) := \{\mathbf{x} \mid (\mathbf{x} - x_0) \wedge v = 0\}
\tag{4.17}
$$

$$
[(x_1\sigma_1 + x_2\sigma_2 + x_3\sigma_3 + x_4\sigma_4) - (\sigma_1 + 2\sigma_2 + 3\sigma_3 + 4\sigma_4)] \\
\wedge (\sigma_1 + \sigma_2 + \sigma_3 + \sigma_4) = 0
$$

and

$$
[(x_1 - 1)\sigma_1 + (x_2 - 2)\sigma_2 + (x_3 - 3)\sigma_3 + (x_4 - 4)\sigma_4] \wedge (\sigma_1 + \sigma_2 + \sigma_3 + \sigma_4) = 0
$$

The outer product $(x - x_0) \wedge v = \frac{1}{2}[(x - x_0)v - v(x - x_0)]$,

$$[(x_1 - 1)\sigma_1 + (x_2 - 2)\sigma_2 + (x_3 - 3)\sigma_3 + (x_4 - 4)\sigma_4](\sigma_1 + \sigma_2 + \sigma_3 + \sigma_4) =$$
$$(\sigma_1 + \sigma_2 + \sigma_3 + \sigma_4 - 10) + (x_1 - x_2 + 1)\sigma_{12} + (x_1 - x_3 + 2)\sigma_{13} + (x_1 - x_4 + 3)$$
$$\sigma_{14} + (x_2 - x_3 + 1)\sigma_{23} + (x_2 - x_4 + 2)\sigma_{24} + (x_3 - x_4 + 1)\sigma_{34}$$

and

$$(\sigma_1 + \sigma_2 + \sigma_3 + \sigma_4)[(x_1 - 1)\sigma_1 + (x_2 - 2)\sigma_2 + (x_3 - 3)\sigma_3 + (x_4 - 4)\sigma_4] =$$
$$(\sigma_1 + \sigma_2 + \sigma_3 + \sigma_4 - 10) - [(x_1 - x_2 + 1)\sigma_{12} + (x_1 - x_3 + 2)\sigma_{13} + (x_1 - x_4 + 3)$$
$$\sigma_{14} + (x_2 - x_3 + 1)\sigma_{23} + (x_2 - x_4 + 2)\sigma_{24} + (x_3 - x_4 + 1)\sigma_{34}]$$

$$x_1 + x_2 + x_3 + x_4 - 10 = 0$$

From Eq. 4.18, the system has multiple solutions $x_1 + x_2 + x_3 + x_4 = 10$, so the points $(x_1, 10 - x_1 - x_3 - x_4, 10 - x_2 - x_4, 10 - x_1 - x_2 - x_3)$ are the solution. Note that the points $(1, 2, 3, 4)$ meet the line $L_{x_0}(v)$.

## 4.6.  Analytical Representation of a Plane in $\mathbb{R}^n$

**Definition 4.13.** Given two **vectors** $v$ and $u$, and a point $x_0$ in space $\mathbb{R}^n$, what is the equation of the plane passing through the point $x_0$ over the plane generated by the vectors $v$ and $u$? (Case adapted with permission of the author [1, 22]). The plane $P_{x_0}(u \wedge v)$ is given by

$$P_{x_0}(u \wedge v) := \{\mathbf{x} \mid (\mathbf{x} - x_0) \wedge (u \wedge v) = 0\} \tag{4.18}$$

Remark 4.6. Note that the oriented plane is defined by $u \wedge v$.

Example 4.22. Given two vectors $v$ and $u$, and a point $x_0$ in space $\mathbb{R}^n$, what is the equation of the plane passing through the point $x_0 = (1, 2, 3, 4)$ in the plane generated by the vectors $v = (1, 0, 0, 0)$ and $u = (0, 1, 0, 0)$?

**Solution 4.22.** The plane $P_{x_0}(u, v)$ is given by

$$P_{x_0 = (1,2,3,4)}(u \wedge v) := \{\mathbf{x} \mid (\mathbf{x} - x_0) \wedge (u \wedge v) = 0\} \tag{4.19}$$

$$[(x_1\sigma_1 + x_2\sigma_2 + x_3\sigma_3 + x_4\sigma_4) - (\sigma_1 + 2\sigma_2 + 3\sigma_3 + 4\sigma_4)] \wedge (\sigma_{12}) = 0$$
$$[(x_1 - 1)\sigma_1 + (x_2 - 2)\sigma_2 + (x_3 - 3)\sigma_3 + (x_4 - 4)\sigma_4] \wedge (\sigma_{12}) = 0 \tag{4.20}$$

From Eq. 4.20, $[(x_1 - 1)\sigma_1 + (x_2 - 2)\sigma_2 + (x_3 - 3)\sigma_3 + (x_4 - 4)\sigma_4][\sigma_{12}] - [\sigma_{12}][(x_1 - 1)\sigma_1 + (x_2 - 2)\sigma_2 + (x_3 - 3)\sigma_3] = 0$, so the points $(2 - x_2, 1 - x_1, 3)$ are the solution. Note that the points $(1, 2, 3)$ meet the plane $P_{x_0 = (1,2,3)}(u \wedge v)$.

## 4.7. Remarks

Note that the outer product $a \wedge b \wedge c$ is neither a scalar nor a vector. In fact, it is the region in space $\mathbb{R}^n$ spanned by the vectors $a$, $b$, and $c$ that represent a region with a volume and an orientation; interchange $a$, $b$, and $c$ and you will change the sign of $a \wedge b \wedge c$ [2, 9, 20, 22, 23, 24, 26]. With this chapter, we have finished the study of this algebra in the $\mathbb{R}^n$ space.

In the following chapters (Chaps. 5-7), we will show how the **Geometric algebra** or **Grassmann algebra** is used in the generalization of the $n$ dimensional space and the introduction of the operatives related with differentiation and integration.

## 4.8. Exercises

**Exercise 4.1.** Provide some exercises of elements of $\mathbb{G}_n$.

**Exercise 4.2.** Let two elements $a = (-1, 1, 1, -1, 2)$ and $b = (2, 3, -3, 1, -1, 3) \in \mathbb{G}_n$ [1]. (i) Express these elements using the orthonormal basis. (ii) Provide examples of a plane and a line.

**Exercise 4.3.** Consider two elements $a = \sigma_{12345}$ and $b = \sigma_1 \in \mathbb{G}_n$. Obtain the outer product $a \wedge b$.

**Exercise 4.4.** Let the vectors $a = -\sigma_{12345}$, and $b = \sigma_1 \in \mathbb{G}_n$. Determine its inner product.

**Exercise 4.5.** Consider elements $a = \sigma_{1234}, b = -\sigma_{5678} \in \mathbb{G}_n$. (i) Obtain the geometric products $ab$ and $ba$. (ii) Determine $a \cdot b$. (iii) Determine $a \wedge b$.

**Exercise 4.6.** Let three elements $a = \sigma_{1234}, b = \sigma_1 - \sigma_2$, and $c = \sigma_1 + \sigma_{123} \in \mathbb{G}_n$ [1]. (i) Determine $a(b + c)$. (ii) Determine $ab + ac$. (iii) Is the distributivity fulfilled?

**Exercise 4.7.** Let three elements $a = \sigma_{1234}, b = \sigma_1$, and $c = -\sigma_{12} \in \mathbb{G}_n$. (i) Determine $a(b + c)$ and $(b + c)a$. (ii) Determine $a \wedge (b + c)$. (iii) Determine $a \wedge b$ and $a \wedge c$. (iv) Is the distributivity over the outer product fulfilled?

**Exercise 4.8.** Let an element $a = 1 + \sigma_1 + \sigma_1 + \cdots + \sigma_n \in \mathbb{G}_n$. Obtain the element $a^{-1}$.

**Exercise 4.9.** Let an element $a = \sigma_{123456} \in \mathbb{G}_n$. Obtain the reverse of element $a$.

**Exercise 4.10.** Let an element $a = 1 + 2\sigma_1\sigma_2 - \sigma_{123456} \in \mathbb{G}_n$. Obtain its blades.

**Exercise 4.11.** Let two elements $a = \sigma_1 + \sigma_2 + \cdots + \sigma_n$, and $b = \sigma_1 \in \mathbb{G}_n$. Obtain the dual of $a \wedge b$.

**Exercise 4.12.** Let an element $a = \sigma_1 + \sigma_2 + \cdots + \sigma_n \in \mathbb{G}_n$. Obtain the norm of element $a$.

**Exercise 4.13.** Consider three elements $a = \sigma_{1234}, b = \sigma_{12345}$ and $c = \sigma_{123456} \in \mathbb{G}_n$ where $\alpha \in \mathbb{R}$. (i) Determine $a(bc)$. (ii) Determine $(ab)c$. (iii) Is the associativity fulfilled?

**Exercise 4.14.** Let an element $a = \sigma_{1234} \in \mathbb{G}_n$ [1]. (i) Determine $Ia$. (ii) Determine $aI$. (iii) Explain (i) and (ii).

**Exercise 4.15.** Let an element $a = \sigma_{1234} \in \mathbb{G}_n$ [1]. (i) Determine $IIa$. (ii) Determine $aII$. (iii) Explain (i) and (ii).

**Exercise 4.16.** Hallar el vector $y$ para el vector $x = \sigma_1 + 2\sigma_2 + \cdots + n\sigma_n$. Sobre el vector $u = \sigma_1 + \sigma_2 + \cdots + \sigma_n$.

**Exercise 4.17.** Given a vector $v$ and a point $x_0$ in the plane $\mathbb{R}^5$. What is the equation of the **line** passing through the point $x_0 = (1, 1, 1, 1)$ in the direction of the vector $v = (1, 0, 0, 1)$?

**Exercise 4.18.** Given two vectors $v$ and $u$, and a point $x_0$ in the space $\mathbb{R}^5$. What is the equation of the **plane** passing through point $x_0 = (2, 1, 1, 1, 1)$ in the plane generated by the vectors $v = (0, 1, 0, 1, 0)$ and $u = (1, 1, 1, 1, 1)$?

**CHAPTER 5**

# Differentiation

**Carlos Polanco**

Faculty of Sciences, Universidad Nacional Autónoma de México, México

**Keywords**: $0-$Forms, $1-$Forms, $2-$Forms, 3.Forms, $k-$Forms, $d\eta$, $df$, $dx_i$, $dx_i \wedge dx_j$, $dx_i \wedge dx_j \wedge dx_k$, $dx_i \wedge dx_j \wedge dx_k \wedge dx_l$, $dx_i \wedge dx_j \wedge \cdots \wedge dx_n$, $dw$, $d\eta$, $d(w \wedge \eta)$, derivative of $0-$form, derivative of $1-$form, derivative of $2-$form, derivative of $3-$form, derivative of $k-$form, differential forms, divergence, exterior derivative, function $w$, function $\eta$, geometric product, geometric product, gradient, inner product, outer product, rotational, tangent line, tangent plane

## 5.1.  Differential of a Function

Informally, an approximation to the definition of a differential of the function $f : \mathbb{R} \to \mathbb{R}$ is $dy = f'(x)dx$. If $dy < 0$, then $(dy)^2$ is negligible, i.e. $(dy)^2 \approx 0$ [11]. This assumption is useful to obtain the **derivative**, or exterior derivative, of an **outer product**.

If we substitute the elements $\sigma_i$ in **orthonormal basis** of the **geometric product** by the **differentials** $dx_i$ and consider $(dy)^2 \approx 0$ (Prop. 5.1), then we can define the **families** (of both real-valued functions and vector-valued functions), whose basis are formed by $dx_i$ that act on a **tangent plane**.

This type of **function families** are known as **Differential forms**.

Remark 5.1. **The equivalent is $dx_i \wedge dx_j$, $dx_i dx_j$, and $dx_{ij}$.**

## 5.2. Differential Forms

### 5.2.1. 0−Forms

A 0−form is any differentiable real-valued function $f(x)$ defined to assign a unique **real number** to a point, i.e. a 0-form is the measure of a flux over a point in an infinitesimal 0−region [27].

**Definition 5.1.** A 0-form in $\mathbb{R}^n$ is a **differentiable real-valued function** $w_0$ (Eq. 5.1) [4].

$$w_0 = f : \mathbb{R}^n \to \mathbb{R} \tag{5.1}$$

Example 5.1. Determine the product and sum of the functions $w_{01}(x,y) = e^x + 3y$ and $w_{02}(x,y) = x - y$.

**Solution 5.1.** (i) $w_0(x,y) = w_{01}(x,y) + w_{02}(x,y) = e^x + 3y + x - y = e^x + 2y + x$.
(ii) $w_0(x,y) = w_{01}(x,y)w_{02}(x,y) = (e^x + 3y)(x - y) = xe^x - ye^x + 3yx - 3y^2$.

Example 5.2. Determine the product and sum of the functions $w_{01}(x) = \sin x$ and $w_{02}(x) = \cos x$.

**Solution 5.2.** (i) $w_0(x) = w_{01}(x) + w_{02}(x) = \sin x + \cos x$. (ii) $w_0(x) = w_{01}w_{02}(x) = \sin x \cos x$.

### 5.2.2. 1−Forms

A 1−form is any differentiable vector-valued function $f(x)$ defined to assign a unique **real number** to an oriented curve, i.e. a 1-form is the measure of a flux over an oriented curve in an infinitesimal 1−region [27].

**Definition 5.2.** A 1−Form in $\mathbb{R}^n$ is a **vector-valued function** formed by a linear combination of the real-valued functions $f_i : \mathbb{R}^n \to \mathbb{R}$ over an **orthonormal basis**, formed by the differentials $dx_i$ (Eq. 5.2) [4].

$$w_1 = f_1(x_1, \cdots, x_n)\,dx_1 + \cdots + f_n(x_1, \cdots, x_n)\,dx_n \tag{5.2}$$

Example 5.3. Determine the product and the sum of the functions $w_{11}(x,y) = e^x\,dx + 3y\,dy$, $w_{12}(x,y) = x\,dx - y\,dy$, and $w_0(x,y) = xy$.

**Solution 5.3.** (i) $w_{11}(x,y) + w_{12}(x,y) = (e^x + x)\,dx + (3y - y)\,dy$. (ii) $w_0(x,y)w_{11}(x,y) = (xy)e^x\,dx + (xy)3y\,dy = xye^x\,dx + 3xy^2\,dy$.

Example 5.4. Determine the product and the sum of the functions $w_{11}(x) = \sin x\,dx$ and $w_{12}(x) = \cos x\,dx$.

**Solution 5.4.** (i) $w_1(x) = w_{11}(x) + w_{12}(x) = (\sin x + \cos x)\,dx$. (ii) If $w_0(x) = \tan x$ then $w_0(x)w_{11}(x) = \tan x \sin x\,dx$.

Example 5.5. Determine the product and the sum of the functions $w_{11}(x) = \sin x\,dx$ and $w_{12}(x) = \cos x\,dy$.

**Solution 5.5.** (i) $w_1(x) = w_{11}(x) + w_{12}(x) = (\sin x\,dx + \cos x)\,dy$. (ii) If $w_0(x) = \tan x$ then $w_0(x)w_{11}(x) = \tan x \sin x\,dx$.

## 5.2.3.   2−Forms

A 2−form is any differentiable vector-valued function $f(x)$ defined to assign a unique **real number** to an oriented surface, i.e. a 2-form is the measure of a flux in an infinitesimal 2−region [27].

**Definition 5.3.** A 2−Form in $\mathbb{R}^n$, is a **vector-valued function** formed by a linear combination of the real-valued functions $f_i : \mathbb{R}^n \to \mathbb{R}$ over an **orthonormal basis** of the differentials $dx_i \wedge dx_j$ (Eq. 5.3) [4].

$$w_2 = f_1(x_1, \cdots, x_n)\,dx_1 \wedge dx_2 + \cdots + f_n(x_1, \cdots, x_n)\,dx_i \wedge dx_j \qquad (5.3)$$

Example 5.6. Determine the product and the sum of the functions $w_{21}(x,y,z) = e^x\,dx \wedge dy + 3zy\,dx \wedge dz + \cos x\,dy \wedge dz$, $w_{22}(x,y,z) = xz\,dx \wedge dy - y\,dx \wedge dz$, and $w_0(x,y,z) = xy$.

**Solution 5.6.** (i) $w_{21}(x,y,z) + w_{22}(x,y,z) = (e^x + xz)\,dx \wedge dy + (3yz - y)\,dx \wedge dz + \cos x\,dy \wedge dz$. (ii) $w_0(x,y,z)w_{21}(x,y,z) = (xy)e^x\,dx \wedge dy + 3xy^2z\,dx \wedge dz + xy\cos x\,dy \wedge dz$.

Example 5.7. Determine the product and the sum of the functions $w_{21}(x,y) = \sin x\,dx \wedge dy$ and $w_{22}(x,y) = \cos x\,dx \wedge dy$.

**Solution 5.7.** (i) $w_1(x,y) = w_{21}(x,y) + w_{22}(x,y) = (\sin x + \cos x)\,dx \wedge dy$. (ii) If $w_0(x,y) = \tan x$ then $w_0(x,y)w_{21}(x,y) = \tan x \sin x\,dx \wedge dy$.

## 5.2.4.   3−Forms

A 3−form is any differentiable vector-valued function $f(x)$ defined to assigne a unique **real number** to an oriented volume, i.e. a 3-form is the measure of a flux over an oriented volume in an infinitesimal 3−region, it is the measure of a fluid [27].

**Definition 5.4.** A 3−Form in $\mathbb{R}^n$ is a **vector-valued function** formed by a linear combination of the real-valued functions $f_i : \mathbb{R}^n \to \mathbb{R}$ over an **orthonormal basis** of the differentials $dx_i \wedge dx_j \wedge dx_k$ (Eq. 5.4).

$$w_3 = f_1(x_1, \cdots, x_n)\,dx_1 \wedge dx_2 \wedge dx_3 + \cdots + f_n(x_1, \cdots, x_n)\,dx_i \wedge dx_j \wedge dx_k \quad (5.4)$$

Example 5.8. Determine the product and the sum of the functions $w_{31}(x,y,z) = e^x\,dx \wedge dy \wedge dz + 3zy\,dx \wedge dz \wedge dy + \cos x\,dy \wedge dz \wedge dx$, $w_{32}(x,y,z) = xz\,dx \wedge dy \wedge dz - y\,dx \wedge dz \wedge dy$, and $w_0(x,y,z) = xy$.

**Solution 5.8.** (i) $w_{31}(x,y,z) + w_{32}(x,y,z) = (e^x + xz)\,dx \wedge dy \wedge dz + (3yz - y)\,dx \wedge dz \wedge dy + \cos x\,dy \wedge dz \wedge dx$. (ii) $w_0(x,y,z)w_{31}(x,y,z) = (xy)e^x\,dx \wedge dy \wedge dz + 3xy^2z\,dx \wedge dz \wedge dy + xy\cos x\,dy \wedge dz \wedge dx$.

Example 5.9. Determine the product and the sum of the functions $w_{31}(x,y,z) = \sin x\,dx \wedge dy \wedge dz$ and $w_{32}(x,y,z) = \cos x\,dx \wedge dy \wedge dz$.

**Solution 5.9.** (i) $w_1(x,y,z) = w_{31} + w_{32} = (\sin x + \cos x)\,dx \wedge dy \wedge dz$. (ii) If $w_0(x,y,z) = \tan x$, then $w_0(x,y,z)w_{31}(x,y,z) = \tan x \sin x\,dx \wedge dy \wedge dz$.

## 5.2.5.   $k-$Forms

A $k-$form is any differentiable vector-valued function $f(x)$ defined to assign a unique **real number** to an oriented $k-$volume, i.e. a $k$-form is the measure of a flux over a $k-$volume in an infinitesimal $k-$region [27].

**Definition 5.5.** A $k-$Form in $\mathbb{R}^n$ is a **vector-valued function** formed by a linear combination of the real-valued functions $f_i : \mathbb{R}^n \to \mathbb{R}$ over an **orthonormal basis** of the differentials $dx_i \wedge dx_j \wedge \cdots \wedge dx_n$ (Eq. 5.5).

$$w_n = f_1(x_1, \cdots, x_n)\,dx_1 \wedge dx_2 \wedge dx_3 + \cdots + f_n(x_1, \cdots, x_n)\,dx_i \wedge dx_j \wedge dx_k \quad (5.5)$$

Example 5.10. Determine the product and the sum of the functions $w_{41}(x_1,x_2,x_3,x_4) = e^x\,dx_1 \wedge dx_2 + 3zy\,dx_1 \wedge dx_3 + \cos x\,dx_2 \wedge dx_3$, $w_{42}(x_1,x_2,x_3,x_4) = xz\,dx_1 \wedge dx_2 - y\,dx_1 \wedge dx_3$, and $w_0(x_1,x_2,x_3,x_4) = xy$.

**Solution 5.10.** (i) $w_{41}(x_1,x_2,x_3,x_4) + w_{42}(x_1,x_2,x_3,x_4) = (e^x + xz)\,dx_1 \wedge dx_2 + (3yz - y)\,dx_1 \wedge dx_3 + \cos x\,dx_2 \wedge dx_3$. (ii) $w_0(x_1,x_2,x_3,x_4)w_{41}(x_1,x_2,x_3,x_4) = (xy)e^x\,dx_1 \wedge dx_2 + 3xy^2z\,dx_1 \wedge dx_3 + xy\cos x\,dx_2 \wedge dx_3$.

Example 5.11. Determine the product and the sum of the functions $w_{41}(x_1,x_2,x_3,x_4) = \sin x\,dx_1$ and $w_{42}(x_1,x_2,x_3,x_4) = \cos x\,dx_1$.

**Solution 5.11.** (i) $w_1(x_1,x_2,x_3,x_4) = w_{41}(x,x_2,x_3,x_4) + w_{42}(x_1,x_2,x_3,x_4) = (\sin x + \cos x)\,dx_1$. (ii) If $w_0(x_1,x_2,x_3,x_4) = \tan x$ then $w_0(x_1,x_2,x_3,x_4)w_{41}(x_1,x_2,x_3,x_4) = \tan x \sin x\,dx_1$.

## 5.3.   **Differentiation of Forms**

**Definition 5.6.** The derivative, or **exterior derivative**, of a $k-$form function is a $(k+1)-$form function (Eq. 5.6).

$$dw = \frac{\partial f}{\partial x_1}dx_1 + \frac{\partial f}{\partial x_2}dx_2 + \cdots \frac{\partial f}{\partial x_n}dx_n \qquad (5.6)$$

## 5.3.1. Derivative of $0-$Forms

A $dx_0$ is the measurement of a net flux through the boundary of an infinitesimal $1-$region [27] at a point.

**Definition 5.7.** The derivative [3] of a $0-$form function $w_0$ (Eq. 5.7) is a $1-$form function of $C^1$ class $w_1 = dw_0$ (Eq. 5.8).

$$w_0 = f(x_1, \cdots, x_n) \tag{5.7}$$

$$\begin{aligned} dw_0 &= \sum_{i=1}^{n} \frac{\partial f(x_1, \cdots, x_n)}{\partial x_i} dx_i \\ &= \frac{\partial f(x_1, \cdots, x_n)}{\partial x_1} dx_1 + \cdots + \frac{\partial f(x_1, \cdots, x_n)}{\partial x_n} dx_n \end{aligned} \tag{5.8}$$

The **field associated** to $w_1 = dw_0$ is the **gradient** of the $0-$form $w_0$ (Sect. 5.3.6).

Example 5.12. Let $w_0 = P(x,y,z)$ in $\mathbb{R}^3$ be a 0-form. Determine the derivative $dw_0$.

**Solution 5.12.**

$$dw_0 = \frac{\partial P}{\partial x} dx + \frac{\partial P}{\partial y} dy + \frac{\partial P}{\partial z} dz \tag{5.9}$$

Example 5.13. Let $w_0 = xy$. Determine its derivative $dw_0$.

**Solution 5.13.**

$$\begin{aligned} dw_0 &= \frac{\partial xy}{\partial x} dx + \frac{\partial xy}{\partial y} dy \\ &= ydx + xdy \end{aligned} \tag{5.10}$$

$\square$

## 5.3.2. Derivative of $1-$Forms

A $dx_1$ is the measure of the net flux through the boundary of an infinitesimal $2-$region [27] in an oriented curve.

**Definition 5.8.** The derivative [3] of a $1-$form function $w_1$ (Eq. 5.11) is a $2-$form function of $C^1$ class $w_2 = dw_1$ (Eq. 5.12).

$$w_1 = f_1(x_1, \cdots, x_n) dx \tag{5.11}$$

$$dw_1 = \left( \sum_{i=1}^{n} \frac{\partial f_1(x_1, \cdots, x_n)}{\partial x_i} \, dx_i \right) \wedge dx \qquad (5.12)$$

The **field associated** to $w_2 = dw_1$ is the **rotational** of the $1-$form $w_1$ (Sect. 5.3.7).

Example 5.14. Let $w_1 = P(x,y,z) \, dx + Q(x,y,z) \, dy$ in $\mathbb{R}^3$ be a 1-form. Determine its derivative $dw_1$.

**Solution 5.14.**

$$dw_1 = d(P \wedge dx) + d(Q \wedge dy)$$

$$= \left( \frac{\partial P}{\partial x} dx + \frac{\partial P}{\partial y} dy + \frac{\partial P}{\partial z} dz \right) \wedge dx + \left( \frac{\partial Q}{\partial x} dx + \frac{\partial Q}{\partial y} dy + \frac{\partial Q}{\partial z} dz \right) \wedge dy$$

$$= \left( \frac{\partial P}{\partial x} \right) dx \wedge dx + \left( \frac{\partial P}{\partial y} \right) dy \wedge dx + \left( \frac{\partial P}{\partial z} \right) dz \wedge dx$$
$$+ \left( \frac{\partial Q}{\partial x} \right) dx \wedge dy + \left( \frac{\partial Q}{\partial y} \right) dy \wedge dy + \left( \frac{\partial Q}{\partial z} \right) dz \wedge dy$$

$$= \left( \frac{\partial Q}{\partial x} - \frac{\partial P}{\partial y} \right) dxdy + \frac{\partial P}{\partial z} dzdx - \frac{\partial Q}{\partial z} dydz$$

$$(5.13)$$

Example 5.15. Let $w_1 = xy \, dx + e^z \, dy + x \, dz$ [28]. Determine its $dw_1$.

**Solution 5.15.**

$$dw_1 = d(xy \wedge dx) + d(e^z \wedge dy) + d(x \wedge dz)$$

$$= \left( \frac{\partial xy}{\partial x} dx + \frac{\partial xy}{\partial y} dy + \frac{\partial xy}{\partial z} dz \right) \wedge dx$$
$$+ \left( \frac{\partial e^z}{\partial x} dx + \frac{\partial e^z}{\partial y} dy + \frac{\partial e^z}{\partial z} dz \right) \wedge dy$$
$$+ \left( \frac{\partial x}{\partial x} dx + \frac{\partial x}{\partial y} dy + \frac{\partial x}{\partial z} dz \right) \wedge dz$$

$$(5.14)$$

$$= \left( \frac{\partial xy}{\partial x} \right) dx \wedge dx + \left( \frac{\partial xy}{\partial y} \right) dy \wedge dx + \left( \frac{\partial xy}{\partial z} \right) dz \wedge dx$$
$$+ \left( \frac{\partial e^z}{\partial x} \right) dx \wedge dy + \left( \frac{\partial e^z}{\partial y} \right) dy \wedge dy + \left( \frac{\partial e^z}{\partial z} \right) dz \wedge dy$$
$$= \left( \frac{\partial x}{\partial x} \right) dx \wedge dz + \left( \frac{\partial x}{\partial y} \right) dy \wedge dz + \left( \frac{\partial x}{\partial z} \right) dz \wedge dz$$

$$= -x \, dxdy + dzdx + e^z dzdy$$

□

### 5.3.3. Derivative of $2-$Forms

A $dx_2$ is the measure of the net flux through the boundary of an infinitesimal $3-$region [27] in an oriented surface.

**Definition 5.9.** The derivative [3] of a $2-$form function $w_2$ (Eq. 5.15) is a $3-$form function of $C^1$ class $w_3 = dw_2$ (Eq. 5.16).

$$w_2 = f(x_1, \cdots, x_n)\, dx_j \wedge dx_k \qquad (5.15)$$

$$dw_2 = \left( \sum_{i=1}^{n} \frac{\partial f(x_1, \cdots, x_n)}{\partial x_i}\, dx_i \right) \wedge dx_j \wedge dx_k \qquad (5.16)$$

The **field associated** to $w_3 = dw_2$ is the **divergence** of the $2-$form $w_2$ (Sect 5.3.8).

Example 5.16. Let $w_2 = P(x,y,z)dydz + Q(x,y,z)dzdx + R(x,y,z)dxdy$ in $\mathbb{R}^3$ be a 2-form [29]. Determine its derivative $dw_2$.

**Solution 5.16.**

$$dw_2 = d(Pdydz) + d(Qdzdx) + d(Rdxdy)$$

$$= (\frac{\partial P}{\partial x}dx + \frac{\partial P}{\partial y}dy + \frac{\partial P}{\partial z}dz) \wedge (dy \wedge dz)$$
$$+ (\frac{\partial Q}{\partial x}dx + \frac{\partial Q}{\partial y}dy + \frac{\partial Q}{\partial z}dz) \wedge (dz \wedge dx)$$
$$+ (\frac{\partial R}{\partial x}dx + \frac{\partial R}{\partial y}dy + \frac{\partial R}{\partial z}dz) \wedge (dx \wedge dy)$$

$$= \frac{\partial P}{\partial x}dx(dy \wedge dz) + \frac{\partial P}{\partial y}dy \wedge (dy \wedge dz) + \frac{\partial P}{\partial z}dz(dy \wedge dz)$$
$$+ \frac{\partial Q}{\partial x}dx(dz \wedge dx) + \frac{\partial Q}{\partial y}dy(dz \wedge dx) + \frac{\partial Q}{\partial z}dz(dz \wedge dx) \qquad (5.17)$$
$$+ \frac{\partial R}{\partial x}dx(dx \wedge dy) + \frac{\partial R}{\partial y}dy \wedge (dx \wedge dy) + \frac{\partial R}{\partial z}dz(dx \wedge dy)$$

$$= \frac{\partial P}{\partial x}dxdydz + \frac{\partial P}{\partial y}dydydz + \frac{\partial P}{\partial z}dzdydz$$
$$+ \frac{\partial Q}{\partial x}dxdzdx + \frac{\partial Q}{\partial y}dydzdx + \frac{\partial Q}{\partial z}dzdzdx$$
$$+ \frac{\partial R}{\partial x}dxdxdy + \frac{\partial R}{\partial y}dydxdy + \frac{\partial R}{\partial z}dzdxdy$$

$$= \left( \frac{\partial P}{\partial z} + \frac{\partial Q}{\partial y} + \frac{\partial R}{\partial x} \right) dxdydz$$

Example 5.17. Let $w_{11}(x,y,z) = xdx + 3ydy$ and $w_{12}(x,y,z) = 2ydx + 2xzdy + xydz$ [29]. (i) Determine $w = w_{11} \wedge w_{12}$. (ii) Obtain its $dw$.

**Solution 5.17.** (i)

$$
\begin{aligned}
w &= w_{11} \wedge w_{12} \\
&= \frac{1}{2}(w_{11}w_{12} - w_{12}w_{11}) \\
&= (3zy^2 - 2x^2z)dxdy - x^2ydxdz - 3xy^2dydz
\end{aligned}
\tag{5.18}
$$

(ii)

$$
dw = d((3zy^2 - 2x^2z)dxdy) - d(x^2ydxdz) - d(3xy^2dydz)
$$

$$
\begin{aligned}
&= \left(\frac{\partial 3zy^2 - 2x^2z}{\partial x}dx + \frac{\partial 3zy^2 - 2x^2z}{\partial y}dy + \frac{\partial 3zy^2 - 2x^2z}{\partial z}dz\right)dx \wedge dy \\
&\quad - \left(\frac{\partial x^2y}{\partial x}dx + \frac{\partial x^2y}{\partial y}dy + \frac{\partial x^2y}{\partial z}dz\right)dx \wedge dz \\
&\quad - \left(\frac{\partial 3xy^2}{\partial x}dx + \frac{\partial 3xy^2}{\partial y}dy + \frac{\partial 3xy^2}{\partial z}dz\right)dy \wedge dz
\end{aligned}
$$

$$
= \frac{\partial 3zy^2 - 2x^2z}{\partial z}dx \wedge dy \wedge dz + \frac{\partial x^2y}{\partial y}dx \wedge dy \wedge dz - \frac{\partial 3xy^2}{\partial x}dx \wedge dy \wedge dz
$$

$$
= (3y^2 - 2x^2 - x^2 - 3y^2)\,dx \wedge dy \wedge dz
$$

$$
= -3x^2\,dx \wedge dy \wedge dz
\tag{5.19}
$$

## 5.3.4.   Derivative of $3-$Forms

A $dx_3-$ is a measure of the net flux through the boundary of an infinitesimal $4-$region [27] in an oriented volume.

**Definition 5.10.** The derivative [3] of a $3-$form function $w_3$ (Eq. 5.20) is a $4-$form function of $C^1$ class $w_4 = dw_3$ (Eq. 5.21).

$$
w_3 = f(x_1, \cdots, x_n)\,dx_1 \wedge dx_2 \wedge dx_3
\tag{5.20}
$$

$$
dw_3 = \left(\sum_{i=1}^{n} \frac{\partial f(x_1, \cdots, x_n)}{\partial x_i}dx_i\right) \wedge dx_j \wedge dx_k \wedge dx_l
\tag{5.21}
$$

Example 5.18. Let $w_3 = P(x_1, x_2, x_3, x_4)dx_2dx_3dx_4$ in $\mathbb{R}^4$ be a 3-form [29]. Determine its derivative $dw_3$.

**Solution 5.18.**

$$dw_3 = d(Pdx_4dx_2dx_3dx)$$

$$= \left( \frac{\partial P}{\partial x_1}dx_1 + \frac{\partial P}{\partial x_2}dx_2 + \frac{\partial P}{\partial x_3}dx_3 + \frac{\partial P}{\partial x_4}dx_4 \right) \wedge (dx_2 \wedge dx_3 \wedge dx_4)$$

$$= \frac{\partial P}{\partial x_1}dx_1 \wedge (dx_2 \wedge dx_3 \wedge dx_4) + \frac{\partial P}{\partial x_2}dx_2 \wedge (dx_2 \wedge dx_3 \wedge dx_4)$$
$$+ \frac{\partial P}{\partial x_3}dx_3 \wedge (dx_2 \wedge dx_3 \wedge dx_4) + \frac{\partial P}{\partial x_4}dx_4 \wedge (dx_2 \wedge dx_3 \wedge dx_4) \tag{5.22}$$

$$= \frac{\partial P}{\partial x_1}dx_1dx_2dx_3dx_4 + \frac{\partial P}{\partial x_2}dx_2dx_2dx_3dx_4$$

$$+ \frac{\partial P}{\partial x_3}dx_3dx_2dx_3dx_4 + \frac{\partial P}{\partial x_4}dx_4dx_2dx_3dx_4$$

$$= \frac{\partial P}{\partial l}dxdydzdl$$

## 5.3.5.   Derivative of $k-$Forms

A $dx_k$ is a measure of the net flux through the boundary of an infinitesimal $(k + 1)-$region [27] enclosed in an oriented $k-$volume.

**Definition 5.11.** The derivative of a $k-$form function $w_k$ (Eq. 5.20) is a $(k + 1)-$form function of $C^1$ class $w_{x+1} = dw_k$ (Eq. 5.24).

$$w_k = f(x_1, \cdots, x_n)dx_j \wedge \cdots \wedge dx_k \tag{5.23}$$

$$dw_k = \left( \sum_{i=1}^{n} \frac{\partial f(x_1, \cdots, x_n)}{\partial x_i}dx_i \right) \wedge dx_j \wedge \cdots \wedge dx_k \tag{5.24}$$

Example 5.19. Let $w_k = P(x, \cdots, k)dy \cdots dk$ in $\mathbb{R}^{(k+1)}$ be a $k$-form [29]. Determine its derivative $dw_k$.

**Solution 5.19.**

$$dw_k = d(Pdk \cdots dx)$$

$$= \left( \frac{\partial P}{\partial x} dx + \cdots + \frac{\partial P}{\partial k} dk \right) \wedge (dy \wedge \cdots dk)$$

$$= \frac{\partial P}{\partial x} dx(dy \wedge \cdots \wedge dk) + \cdots + \frac{\partial P}{\partial k} dk(dy \wedge \cdots \wedge dk) \tag{5.25}$$

$$= \frac{\partial P}{\partial x} dx \cdots dk + \frac{\partial P}{\partial y} dy \cdots dk + \frac{\partial P}{\partial z} dz \cdots dk + \cdots + \frac{\partial P}{\partial k} dk \cdots dk$$

$$= \frac{\partial P}{\partial k} dx \cdots dk$$

## 5.3.6.   Gradient Operator Associated to $dw_0$

**Definition 5.12.** The **gradient** operator is associated to the derivative of an $n$ degree $0-$form $dw_0$ (Eq. 5.26).

$$\mathbf{grad}(dw_0) = \left( \frac{\partial f}{\partial x_1}, \cdots, \frac{\partial f}{\partial x_n} \right) \tag{5.26}$$

Example 5.20. Let $w_0 = P(x,y,z)$. (i) Determine its $dw_0$. (ii) Determine the associated **gradient** to $dx_0$.

**Solution 5.20.** (i)

$$dw_0 = \frac{\partial P}{\partial x} dx + \frac{\partial P}{\partial y} dy + \frac{\partial P}{\partial z} dz \tag{5.27}$$

(ii) The associated **gradient** is

$$\mathbf{grad}(f) = \left( \frac{\partial P}{\partial x}, \frac{\partial P}{\partial y}, \frac{\partial P}{\partial z} \right)$$

.

## 5.3.7.   Rotational Operator Associated to $dw_1$

**Definition 5.13.** The **rotational** operator is associated to the derivative of an $n$ degree $1-$form $dw_1$ (Eq. 5.28).

$$\mathbf{rot}(dw_1) = c_1 \frac{\partial f_1}{\partial x_1} dx_i \wedge dx_j + \cdots + c_n \frac{\partial f_n}{\partial x_n} dx_j \wedge dx_k \tag{5.28}$$

Example 5.21. Let $w_1 = P(x,y,z)\,dx + Q(x,y,z)\,dy + R(x,y,z)\,dz$. (i) Determine its $dw_1$. (ii) Determine the associated **rotational** to $dx_1$.

**Solution 5.21.** (i)

$$dw_1 = d(P \wedge dx) + d(Q \wedge dy) + d(R \wedge dz)$$

$$= \left( \frac{\partial P}{\partial x} dx + \frac{\partial P}{\partial y} dy + \frac{\partial P}{\partial z} dz \right) \wedge dx$$
$$+ \left( \frac{\partial Q}{\partial x} dx + \frac{\partial Q}{\partial y} dy + \frac{\partial Q}{\partial z} dz \right) \wedge dy$$
$$+ \left( \frac{\partial R}{\partial x} dx + \frac{\partial R}{\partial y} dy + \frac{\partial R}{\partial z} dz \right) \wedge dz$$

$$= \left( \frac{\partial P}{\partial x} \right) dx \wedge dx + \left( \frac{\partial P}{\partial y} \right) dy \wedge dx + \left( \frac{\partial P}{\partial z} \right) dz \wedge dx$$
$$+ \left( \frac{\partial Q}{\partial x} \right) dx \wedge dx + \left( \frac{\partial Q}{\partial y} \right) dy \wedge dx + \left( \frac{\partial Q}{\partial z} \right) dz \wedge dx$$ (5.29)
$$+ \left( \frac{\partial R}{\partial x} \right) dx \wedge dx + \left( \frac{\partial R}{\partial y} \right) dy \wedge dx + \left( \frac{\partial R}{\partial z} \right) dz \wedge dx$$

$$= \left( \frac{\partial R}{\partial y} - \frac{\partial Q}{\partial z} \right) dydz + \left( \frac{\partial P}{\partial z} - \frac{\partial R}{\partial x} \right) dzdx + \left( \frac{\partial Q}{\partial x} - \frac{\partial P}{\partial y} \right) dxdy$$

(ii) The associated **rotational** is

$$\mathbf{rot}(dw_1) = \left( \frac{\partial R}{\partial y} - \frac{\partial Q}{\partial z} \right) dydz + \left( \frac{\partial P}{\partial z} - \frac{\partial R}{\partial x} \right) dzdx + \left( \frac{\partial Q}{\partial x} - \frac{\partial P}{\partial y} \right) dxdy$$

## 5.3.8.   Divergence Operator Associated to $dw_2$

**Definition 5.14.** The **divergence** operator is associated to the derivative of an $n$ degree 2−form $dw_2$ (Eq. 5.30).

$$\mathbf{div}(dw_2) = c_1 \frac{\partial f}{\partial x} dx_i \wedge dx_j \wedge dx_k$$ (5.30)

Example 5.22. Let $w_2 = P(x,y,z)\,dydz + Q(x,y,z)\,dzdx + R(x,y,z)\,dxdy$. (i) Determine its $dw_2$. (ii) Determine the associated **divergence** to $dx_2$.

**Solution 5.22.** (i) From (Ex. 5.16)

$$dw_2 = d(Pdydz) + d(Qdzdx) + d(Rdxdy)$$

$$= \left( \frac{\partial P}{\partial z} + \frac{\partial Q}{\partial y} + \frac{\partial R}{\partial x} \right) dxdydz$$ (5.31)

(ii) The associated **divergence** is

$$\mathbf{div}(dw_2) = \left( \frac{\partial P}{\partial z} + \frac{\partial Q}{\partial y} + \frac{\partial R}{\partial x} \right) dxdydz$$

## 5.4.  Remarks

The introduction to **differential forms** in this chapter uses the review of **geometric algebra** in 2D (Chap. 2), in 3D (Chap. 3), and in nD (Chap. 5).

In (Chap. 6) we use the concept of **differentation** of **differential forms** to introduce the **integral** operator over two types of integral **line integral** and **surface integral**.

It is important that the reader reviews and solves all exercises in this chapter, to get familiar with the **differentiation** operator studied here.

## 5.5. Exercises

**Exercise 5.1.** Give examples of the differential forms $w_0$, $w_1$, $w_2$, $w_3$, and $w_4$. Explain the type of each example.

**Exercise 5.2.** Let $w_0(x,y,z) = e^{x^2yz}$. Determine $dw_0$.

**Exercise 5.3.** Let $w_0(x,y,z) = e^{x^2yz}dx + \sin xyz dy$. Determine $dw_1$.

**Exercise 5.4.** Find the $dw_0$ of $w_0(x,y) = x^2y + y^3$.

**Exercise 5.5.** Find the $dw_2$ of $w_2(x,y,z) = (x^3y + y^3)dydz$.

**Exercise 5.6.** Find the $d(dw_2)$ of $w_2(x,y,z) = \dfrac{-x}{x^2y + y^2}dxdy$.

**Exercise 5.7.** Let $w_{11}(x,y,z) = xdx + yzdy + x^3ydz$ and $w_{12}(x,y,z) = xydz$. (i) Compute $d(w_{11})$. (ii) Compute $d(w_{12})$. (iii) Compute $w_{11} \wedge w_{12}$. (iv) Compute $d(w_{11} \wedge w_{12})$. (v) Explain (iv).

**Exercise 5.8.** Compute the outer product of $w_{11}(x,y) = 3dx + dy$, and $w_{12}(x,y) = e^x dx + 2dy$.

**Exercise 5.9.** Compute $(-r\sin\theta d\theta + \cos\theta dr) \wedge (r\cos\theta d\theta + \sin\theta dr)$[1, 17].

**Exercise 5.10.** Let $w_4(x_1,x_2,x_3,x_4) = x_1 + x_3$. (i) Compute $w_4 \wedge w_4$. (ii) Compute $d(w_4 \wedge w_4)$.

**CHAPTER 6**

# Integration

**Carlos Polanco**

Faculty of Sciences, Universidad Nacional Autónoma de México, México

**Abstract** This chapter intends to be a survey on the integration of differential forms. Here, the $0-$Form, $1-$Form, $2-$Form, $3-$Form and $k-$Form integrals are defined. These forms are reviewed using mapping, in particular, the cases that give rise to Simple Riemann integral, Double Riemann integral, Triple Riemann Integral, and the Line and Surface integrals. The latter two are defined in (Chap. 1), using Heaviside-Gibbs algebra.

**Keywords**: Double Riemann integral case, integral of $2-$Form, integral of $3-$Form, integration of $0-$Form, integration of $1-$Form, integration of $k-$Form, line integral, mapping a $1-$Form case, mapping a $2-$Form case, mapping a $3-$Form case, simple Riemann integral case, surface integral, triple Riemann integral case

## 6.1.  Preliminaries

It is important to review all the examples for **line integral** and **surface integral** in this chapter, they are the same in Chap. 1 but with the method studied here, so the reader can compare both methods.

Remark 6.1. **Note that from now on an equivalent notation will be $\sigma_i \wedge \sigma_j$ and $\sigma_i \sigma_j$.**

## 6.2.  Integration of $0-$Forms

The integral of a $0-$form is a **real number** that represents the effect or the total flux that a real-valued function $f(x)$ has over all the points on an interval of the domain of the function [4].

**Definition 6.1.** The integral of a differential form $w_0(x)$ in $\mathbb{R}^n$ over an interval $D \in \mathbb{R}^n$ is represented by (Eq. 6.1).

$$\int_D w_0 = f(a_n) - f(a_1) \tag{6.1}$$

Example 6.1. Let $w_0 = x^3 + 2x$ be a 0-form. Determine the integral on the interval $[1,2]$.

**Solution 6.1.** $\int_D w_0 = \int_1^2 x^3 + 2x = (2^3 + 4) - (1^3 + 2) = 12 - 3 = 9$.

Example 6.2. Let $w_0 = x^3 + 2xy - z$ be a 0-form in $\mathbb{R}^3$. Determine the integral on the interval $[1,2] \times [3,4] \times [5,6]$.

**Solution 6.2.** $\int_D w_0 = \int_1^2 \int_3^4 \int_5^6 x^3 + 2xy - z = (2^3 + 16 - 6) - (1^3 + 6 - 5) = 18 - 2 = 16$.

## 6.3. Integration of $1-$Forms

The integral of a $1-$form is a **real number** that represents the effect or the total flux, that a vector-valued function $f(x)$ has over the oriented curve on an interval of the domain of that function [4].

**Definition 6.2.** The integral of a differential form $w_1$ in $\mathbb{R}^n$ over an interval $D \in \mathbb{R}^n$ is represented by Eq. 6.2.

$$\int_D w_1 = \int_{a_1}^{a_n} f_1(x_1, \cdots, x_n)\, dx_1 + \cdots + f_n(x_1, \cdots, x_n)\, dx_n\, dD \tag{6.2}$$

Here, we show the general method to solve this integral (Def. 6.3.1) and a particular case (Def. 6.4) that is used in **Heaviside-Gibbs algebra**.

## 6.3.1. Mapping a $1-$Form Case

**Definition 6.3.** The integral of a differential form $w_1$ on $\mathbb{R}^n$ over the **curve** $c(t)$ : $[t_1, t_n] \subset \mathbb{R} \to \mathbb{R}^n$ is represented by Eq. 6.3.

$$\int_D w_1 = \int_{t_1}^{t_n} f_1(c(t)) \frac{dx_1}{dt} + \cdots + f_n(c(t)) \frac{dx_n}{dt}\, dt \tag{6.3}$$

Example 6.3. Let $w_1 = x^3 dx + 2xy dy - z dz$ be a 1-form in $\mathbb{R}^3$. Determine the integral over the curve $c(t) = (t, t^2, t^3)$ on the interval $[0,1]$.

**Solution 6.3.** $\int_D w_1 = \int_0^1 x^3 dx + 2xy dy - z dz = \int_0^1 t^3 (t)'_t + 2t^3 (t^2)'_t - t(t^3)'_t\, dt = \int_0^1 -2t^3 + 4t^4\, dt = \frac{3}{10}$.

This integral (Eq. 6.3) is known as **Line integral** in **Heaviside-Gibbs algebra** (Def. 1.3.3). Below, we will see the same example but solved with **Heaviside-Gibbs algebra** and **Grassmann algebra**.

**Example 6.4.** A particle moves along the oriented trajectory $T(t) = (t, t^4)$, $t \in [0, \pi]$ and the force is represented by the vector field $F(x, y) = (-yx, \sin x)$. Compute the work done by the force field on a particle that moves along curve $T$.

**Solution 6.4.** $\oint_T F \circ T(t) \cdot T'(t) \, dt = \int_0^\pi (-t^5, \sin t) \cdot (1, 4t^3) \, dt = \int_0^\pi -t^5 + 4t^3$

$\sin t \, dt = -\dfrac{\pi^6 - 24\pi^3 + 144\pi}{6}.$

**Example 6.5.** Let $w_1 = -yx dx + \sin x dy$ be a 1-form in $\mathbb{R}^2$. Determine the integral over the curve $c(t) = (t, t^4)$ on the interval $[0, \pi]$.

**Solution 6.5.** $\int_D w_1 = \int_0^\pi -yx dx + \sin x dy = \int_0^\pi -t^5 (t)'_t + \sin t (t^4)'_t \, dt = \int_0^\pi -t^5$

$+ 4t^3 \sin t \, dt = -\dfrac{\pi^6 - 24\pi^3 + 144\pi}{6}.$

## 6.3.2. Simple Riemann Integral Case

**Definition 6.4.** An integral of a differential form $w_1$ in $\mathbb{R}$ is equivalent to a Simple Riemann Integral (Eq. 6.4).

$$\int_D w_1 = \int_{a_1}^{a_n} f(x_1) \, dx_1 \qquad (6.4)$$

**Example 6.6.** Let $w_1 = x^3 + 2x dx$ be a 1-form. Determine the integral on the interval $[1, 2]$.

**Solution 6.6.** $\int_D w_1 = \int_1^2 x^3 + 2x dx = \left[ \dfrac{x^4}{4} + x^2 \right]_1^2 = \dfrac{27}{4}.$

## 6.4. Integration of $2-$Forms

The integral of a $2-$form is a **real number** that represents the total effect or flux that a vector-valued function $f(x)$ has over an oriented surface on an interval of the domain of the function [4].

**Definition 6.5.** The integral of a differential form $w_2$ in $\mathbb{R}^n$ over a region $D \in \mathbb{R}^n$, is represented by Eq. 6.5.

$$\int_D w_2 = \int_{a_1}^{a_k} \int_{a_{k+1}}^{a_n} f_1(x_1, \cdots, x_n) \, dx_1 dx_2 + \cdots +$$

$$f_n(x_1, \cdots, x_n) \, dx_n dx_1 \, dD \quad (6.5)$$

Below, we show the general method to solve this integral (Def. 6.4.1) and a particular case of it (Def. 6.7), using the **Heaviside-Gibbs algebra**.

### 6.4.1.  Mapping a $2-$Form Case

**Definition 6.6.** The integral of a differential form $w_2$ in $\mathbb{R}^n$ over the **surface** $c(u,v) : [u_1,u_n] \times [v_1,v_n] \subset \mathbb{R}^2 \to \mathbb{R}^n$ is represented by Eq. 6.6.

$$\int_D w_2 = \int_{u_{11}}^{u_{1n}} \int_{u_{21}}^{u_{2n}} f_1(c(u_1,u_2)) \frac{\partial(x_1,x_2)}{\partial(u_1,u_2)} + \cdots + f_n(c(u_1,u_2)) \frac{\partial(x_{n-1},x_n)}{\partial(u_1,u_2)} \, du_1 \, du_2$$
(6.6)

where

$$\frac{\partial(x_1,x_2)}{\partial(u_1,u_2)} = \begin{vmatrix} \dfrac{\partial x_1}{\partial u_1} & \dfrac{\partial x_1}{\partial u_2} \\ \dfrac{\partial x_2}{\partial u_1} & \dfrac{\partial x_2}{\partial u_2} \end{vmatrix} \cdots \frac{\partial(x_{n-1},x_n)}{\partial(u_1,u_2)} = \begin{vmatrix} \dfrac{\partial x_{n-1}}{\partial u_1} & \dfrac{\partial x_{n-1}}{\partial u_2} \\ \dfrac{\partial x_n}{\partial u_1} & \dfrac{\partial x_n}{\partial u_2} \end{vmatrix}$$
(6.7)

Remark 6.2. Note that $\dfrac{\partial T}{\partial r} \times \dfrac{\partial T}{\partial \theta} = \left( \dfrac{\partial(y,z)}{\partial(r,\theta)}, \dfrac{\partial(z,x)}{\partial(r,\theta)}, \dfrac{\partial(x,y)}{\partial(r,\theta)} \right)$ (Def. 5.3.2).

Example 6.7. Let $w_2 = -y dx dy + x dy dz$ be a 2-form in $\mathbb{R}^3$. Determine the integral over the surface $c(r,\theta) = (r\cos\theta, r\sin\theta, \theta)$ on the interval $[0,1] \times [0,2\pi]$.

**Solution 6.7.** $\displaystyle \int_D w_2 = \int_0^1 \int_0^{2\pi} [-y dx dy + x dy dz] \, d\theta \, dr = \int_0^1 \int_0^{2\pi} \left[ -r\sin\theta \frac{\partial(x,y)}{\partial(r,\theta)} \right.$

$\displaystyle \left. + r\cos\theta \frac{\partial(y,z)}{\partial(r,\theta)} \right] d\theta \, dr = \int_0^1 \int_0^{2\pi} -r^2 \sin\theta + r\cos\theta \sin\theta \, d\theta \, dr = 0.$

Note 6.1. $\dfrac{\partial(x,y)}{\partial(r,\theta)} = r$, and $\dfrac{\partial(y,z)}{\partial(r,\theta)} = \sin\theta$.

This integral (Eq. 6.6) is known as **Surface integral** in **Heaviside-Gibbs algebra** (Def. 1.3.4). Below, we show the same example solved in **Heaviside-Gibbs algebra** and **Grassmann algebra**.

Example 6.8. Let the external side surface of a solid circle with radius 1, height 1, and the vector field $F(x,y,z) = (x,y,z)$. Compute the surface integral of the vector function over the vector field.

**Solution 6.8.** Using $T(r,\theta) = (r\cos\theta, r\sin\theta, 1)$ with $\theta \in [0,2\pi]$, $r \in [0,1]$.

$$\oiint_S F \circ T(r,\theta) \cdot \eta(r,\theta) \, dS = \int_{\theta_1}^{\theta_2} \int_{r_1}^{r_2} F(T(r,\theta)) \cdot \frac{\partial T}{\partial r} \times \frac{\partial T}{\partial \theta} \, dr \, d\theta$$

$$= \int_0^{2\pi} \int_0^1 F(T(r,\theta)) \cdot \frac{\partial T}{\partial r} \times \frac{\partial T}{\partial \theta} \, dr \, d\theta$$
(6.8)

$$= \int_0^{2\pi} \int_0^1 (r\cos\theta, r\sin\theta, 1) \cdot (0,0,r) \, dr \, d\theta$$

$$= \pi.$$

Example 6.9. Let $w_2 = x dx dy + y dy dz + z dz dx$ be a 2-form in $\mathbb{R}^3$. Determine the integral over the curve $c(r,\theta) = (\cos\theta, \sin\theta, 1)$ with $\theta \in [0,2\pi]$, $r \in [0,1]$.

**Solution 6.9.** $\int_D w_2 = \int_0^1 \int_0^{2\pi} [z\,dxdy + x\,dydz + y\,dzdx]\,d\theta dr = \int_0^1 \int_0^{2\pi} \left[ 1\frac{\partial(x,y)}{\partial(r,\theta)} \right.$

$\left. + r\cos x\frac{\partial(y,z)}{\partial(r,\theta)} + r\sin x\frac{\partial(z,x)}{\partial(r,\theta)} \right] d\theta dr = \int_0^1 \int_0^{2\pi} r\,d\theta dr = \pi.$

**Note 6.2.** $\frac{\partial(x,y)}{\partial(r,\theta)} = r$, $\frac{\partial(y,z)}{\partial(r,\theta)} = 0$, and $\frac{\partial(z,x)}{\partial(r,\theta)} = 0$.

## 6.4.2.   Double Riemann Integral Case

**Definition 6.7.** An integral of a differential form $w_2$ in $\mathbb{R}^2$ is equivalent to a Double Riemann Integral (Eq. 6.9).

$$\iint_D w_2 = \int_{a_1}^{a_k} \int_{a_{k+1}}^{a_n} f(x_1,x_2)\,d_1 dx_2 \tag{6.9}$$

Example 6.10. Let $w_2 = x^3 + 2xy\,dxdy$ be a 2-form. Determine the integral on the interval $[1,2] \times [3,4]$.

**Solution 6.10.** $\iint_D w_2 = \int_1^2 \int_3^4 x^3 + 2xy\,dxdy = \frac{57}{4}.$

## 6.5.   Integration of $3-$Forms

The integral of a $3-$form is a **real number** that represents the total effect or flux that a vector-valued function $f(x)$ has over the oriented volume, on an interval of the domain of that function [4].

**Definition 6.8.** The integral of a differential form $w_3$ in $\mathbb{R}^n$ over a region $D \in \mathbb{R}^n$, is represented by Eq. 6.10.

$$\int_D w_3 = \int_{a_1}^{a_k} \int_{a_{k+1}}^{a_m} \int_{a_{m+1}}^{a_n} f_1(x_1,\cdots,x_n)\,dx_1 dx_2 dx_3 + \cdots +$$

$$f_n(x_1,\cdots,x_n)\,dx_n dx_2 dx_1\,dD \tag{6.10}$$

Here, we show the general method to solve this integral (Def. 6.5.1) and a particular case (Def. 6.10) using **Heaviside-Gibbs algebra**.

## 6.5.1.   Mapping a $3-$Form Case

**Definition 6.9.** The integral of a differential form $w_3$ in $\mathbb{R}^n$ over the **volume** $c(u_1,u_2,u_3): [u_{11},u_{1n}] \times [u_{21},u_{2n}] \times [u_{31},u_{3n}] \subset \mathbb{R}^3 \to \mathbb{R}^n$, is represented by Eq. 6.11.

$$\int_D w_3 = \int_{u_{31}}^{u_{3n}} \int_{u_{21}}^{u_{2n}} \int_{u_{11}}^{u_{1n}} f_1(c(u_1,u_2,u_3)) \frac{\partial(x_1,x_2,x_3)}{\partial(u_1,u_2,u_3)} + \cdots +$$

$$f_n(c(u_1,u_2,u_3)) \frac{\partial(x_{n-2},x_{n-1},x_n)}{\partial(u_1,u_2,u_3)} du_1\,du_2\,du_3 \quad (6.11)$$

where

$$\frac{\partial(x_1,x_2,x_3)}{\partial(u_1,u_2,u_3)} = \begin{vmatrix} \dfrac{\partial x_1}{\partial u_1} & \dfrac{\partial x_1}{\partial u_2} & \dfrac{\partial x_1}{\partial u_3} \\[2mm] \dfrac{\partial x_2}{\partial u_1} & \dfrac{\partial x_2}{\partial u_2} & \dfrac{\partial x_2}{\partial u_3} \\[2mm] \dfrac{\partial x_3}{\partial u_1} & \dfrac{\partial x_3}{\partial u_2} & \dfrac{\partial x_3}{\partial u_3} \end{vmatrix}, \cdots, \frac{\partial(x_{n-2},x_{n-1},x_n)}{\partial(u_1,u_2,u_3)} = \begin{vmatrix} \dfrac{\partial x_{n-2}}{\partial u_1} & \dfrac{\partial x_{n-2}}{\partial u_2} & \dfrac{\partial x_{n-2}}{\partial u_3} \\[2mm] \dfrac{\partial x_{n-1}}{\partial u_1} & \dfrac{\partial x_{n-1}}{\partial u_2} & \dfrac{\partial x_{n-1}}{\partial u_3} \\[2mm] \dfrac{\partial x_n}{\partial u_1} & \dfrac{\partial x_n}{\partial u_2} & \dfrac{\partial x_n}{\partial u_3} \end{vmatrix}$$
$$(6.12)$$

Example 6.11. Let $w_3 = xyz\,dz\,dy\,dx$ be a 3-form in $\mathbb{R}^3$. Determine the integral over the volume $c(r,\theta,\phi) = (r,\theta,\phi)$ on the interval $[0,2] \times [0,2\pi] \times [0,\pi]$.

**Solution 6.11.**

$$\int_D w_3 = \int_0^2 \int_0^{2\pi} \int_0^{\pi} [xyz\,dz\,dy\,dx]\,dr\,d\theta\,d\phi =$$

$$\int_0^2 \int_0^{2\pi} \int_0^{\pi} r\theta\phi \frac{\partial(z,y,x)}{\partial(r,\theta,\phi)} = 2\pi^4. \quad (6.13)$$

Note 6.3. $\dfrac{\partial(z,y,x)}{\partial(r,\theta,\phi)} = 1.$

## 6.5.2.  Triple Riemann Integral Case

**Definition 6.10.** An integral of a differential form $w_3$ in $\mathbb{R}^3$ is equivalent to a Triple Riemann Integral (Eq. 6.14).

$$\iiint_D w_3 = \int_{a_1}^{a_k} \int_{a_{k+1}}^{a_m} \int_{a_{m+1}}^{a_n} f(x_1,x_2,x_3)\,dx_1 dx_2 dx_3 \quad (6.14)$$

Example 6.12. Let $w_3 = x^3 z + 2xy\,dz\,dy\,dx$ be a 3-form. Determine the integral on the interval $[1,2] \times [3,4] \times [5,6]$.

**Solution 6.12.** $\iiint_D w_3 = \int_1^2 \int_3^4 \int_5^6 x^3 z + 2xy\,dx\,dy = \dfrac{249}{8}.$

## 6.6.  Integration of $k-$Forms

The integral of a $k-$form is a **real number** that represents the total effect or flux that a vector-valued function $f(x)$ has over the oriented $k-$volume, on an interval of the domain of the function [4].

**Definition 6.11.** The integral of a differential form $w_k$ in $\mathbb{R}^n$ over a region $D \in \mathbb{R}^n$, is represented by Eq. 6.15.

$$\int_D w_k = \int_{a_1}^{a_k} \cdots \int_{a_{k+m}}^{a_n} f_1(x_1, \cdots, x_n)\, dx_1 \cdots dx_2 + \cdots +$$
$$f_n(x_1, \cdots, x_n)\, dx_n \cdots dx_1\, dD \quad (6.15)$$

Below, we show a general method to solve this integral (Def. 6.6.1) and a particular case (Def. 6.13 using **Heaviside-Gibbs algebra**.

## 6.6.1.   Mapping a $k-$Form Case

**Definition 6.12.** The integral of a differential form $w_k$ in $\mathbb{R}^n$ over the $k-$**volume** $c(u_1, \cdots, u_k) : [u_{11}, u_{1n}] \times \cdots \times [u_{k1}, u_{kn}] \subset \mathbb{R}^k \to \mathbb{R}^n$, is represented by Eq. 6.16.

$$\int_D w_k = \int_{u_{k1}}^{u_{kn}} \cdots \int_{u_{11}}^{u_{1n}} f_1\big(c(u_1, \cdots, u_k)\big) \frac{\partial(x_1, \cdots, x_n)}{\partial(u_1, \cdots, u_k)} + \cdots +$$
$$f_n\big(c(u_1, \cdots, u_k)\big) \frac{\partial(x_{n-k}, \cdots, x_n)}{\partial(u_1, \cdots, u_k)}\, du_1 \cdots du_k \quad (6.16)$$

where

$$\frac{\partial(x_1, \cdots, x_n)}{\partial(u_1, \cdots, u_k)} = \begin{vmatrix} \dfrac{\partial x_1}{\partial u_1} & \cdots & \dfrac{\partial x_1}{\partial u_k} \\ \vdots & \ddots & \vdots \\ \dfrac{\partial x_3}{\partial u_1} & \cdots & \dfrac{\partial x_3}{\partial u_k} \end{vmatrix}, \cdots, \frac{\partial(x_{n-k}, \cdots, x_n)}{\partial(u_1, \cdots, u_k)} = \begin{vmatrix} \dfrac{\partial x_{n-2}}{\partial u_1} & \cdots & \dfrac{\partial x_{n-2}}{\partial u_3} \\ \vdots & \ddots & \vdots \\ \dfrac{\partial x_n}{\partial u_1} & \cdots & \dfrac{\partial x_n}{\partial u_3} \end{vmatrix}$$
$$(6.17)$$

**Example 6.13.** Let $w_4 = x_1 x_2 x_3 x_4\, dx_4 dx_3 dx_2 dx_1$ be a 4-form in $\mathbb{R}^4$. Determine the integral over the $4-$volume $c(u_1, u_2, u_3, u_4) = (u_1, u_2, u_3, u_4)$ on the interval $[0,1] \times [0,1] \times [0,1] \times [0,1]$.

**Solution 6.13.**

$$\int_D w_4 = \int_0^1 \int_0^1 \int_0^1 \int_0^1 [x_1 x_2 x_3 x_4\, dx_4 dx_3 dx_2 dx_1]\, du_1 du_2 du_3 du_4 =$$
$$\int_0^1 \int_0^1 \int_0^1 \int_0^1 \left[ x_1 x_2 x_3 x_4 \frac{\partial(x_4, x_3 x_2 x_1)}{\partial(u_4, u_3, u_2, u_1)} \right] du_4 du_3 du_2 du_1 = \frac{1}{16}. \quad (6.18)$$

**Note 6.4.** $\dfrac{\partial(x_4, x_3, x_2, x_1)}{\partial(u_4, u_3, u_2, u_1)} = 1.$

## 6.6.2.   $k-$Riemann Integral Case

**Definition 6.13.** An integral of a differential form $w_k$ in $\mathbb{R}^k$ is equivalent to a $k-$Riemann Integral (Eq. 6.19).

$$\int \cdots \int_D w_k = \int_{a_1}^{a_k} \cdots \int_{a_{k+m}}^{a_n} f(x_1, \cdots, x_k)\, dx_1 \cdots d_k \qquad (6.19)$$

Example 6.14. Let $w_4 = x_1 x_2 x_3 x_4\, dx_4\, dx_3\, dx_2\, dx_1$ be a 4-form. Determine the integral on the interval $[1,2] \times [3,4] \times [5,6] \times [7,8]$.

**Solution 6.14.** $\iiiint_D w_4 = \int_1^2 \int_3^4 \int_5^6 \int_7^8 x_1 x_2 x_3 x_4\, dx_4\, dx_3\, dx_2\, dx_1 = \dfrac{3465}{16}.$

## 6.7.   Remarks

The integration of **differential forms** in this chapter, is based on the review of **geometric algebra** in $2D$ space (Chap. 2), 3D space (Chap. 3), and $nD$ space (Chap. 5).

We suggest the reader reviews and solves all the exercises in this chapter to get familiar with the **integration** operator of this algebra.

Note that in some cases, it is not possible to provide an example in **Heaviside-Gibbs algebra**.

In (Chap. 8) the concepts of **differentiation** and **integration** of **differential forms** are used to introduce Green's theorem, using the **line integral** and **double Riemann integral**.

The previous chapters are the basis to properly understand Green's, Stokes', Gauss' theorems, and Calculus Fundamental Theorem.

## 6.8.   Exercises

**Exercise 6.1.** Let $w_0 = 3x^2 + 2x$ be a 0-form. Determine the integral on the interval $[1,3]$.

**Exercise 6.2.** Let $w_0 = x^2 + 2xy - z$ be a 0-form in $\mathbb{R}^3$. Determine the integral on the interval $[1,2] \times [1,2] \times [1,2]$.

**Exercise 6.3.** Let $w_1 = x^4 dx + 3xy dy - z dz$ be a 1-form in $\mathbb{R}^3$. Determine the integral over the curve $c(t) = (t, t^2, t^3)$ on the interval $[0,2]$.

**Exercise 6.4.** A particle moves along the oriented trajectory $T(t) = (t, t^4), t \in [0, 2\pi]$ and the force is represented by the vector field $F(x,y) = (-2yx, \sin x)$. Compute the work done by the force field on a particle that moves along curve $T$.

**Exercise 6.5.** Let $w_1 = -yx dx + \cos x dy$ be a 1-form in $\mathbb{R}^2$. Determine the integral over the curve $c(t) = (t, t^4)$ on the interval $[-\pi, \pi]$.

**Exercise 6.6.** Let $w_1 = x^4 + 2x \cos x dx$ be a 1-form. Determine the integral on the interval $[-\pi, \pi]$.

**Exercise 6.7.** Let $w_2 = -y dx dy + x^2 dy dz$ be a 2-form in $\mathbb{R}^3$. Determine the integral over the surface $c(r, \theta) = (r \cos \theta, r \sin \theta, \theta)$ on the interval $[0,1] \times [0, \frac{\pi}{2}]$.

**Exercise 6.8.** Let the external side surface of a solid circle with radius 2, height 3, and the vector field $F(x,y,z) = (x,y,z)$. Compute the surface integral of the vector function over the vector field.

**Exercise 6.9.** Let $w_2 = x^3 + 2xy dy dx$ be a 2-form. Determine the integral on the interval $[0,1] \times [0,x]$.

**Exercise 6.10.** Let $w_3 = xyz dz dy dx$ be a 3-form in $\mathbb{R}^3$. Determine the integral over the volume $c(r, \theta, \phi) = (r, \theta, \phi)$ on the interval $[0,1] \times [0, \pi] \times [0, \frac{\pi}{2}]$.

**Exercise 6.11.** Let $w_3 = x^4 z + 2xy dz dy dx$ be a 3-form. Determine the integral on the interval $[0,1] \times [0,1] \times [0,1]$.

**Exercise 6.12.** Let $w_4 = x_1 x_2 x_3 x_4^2 dx_4 dx_3 dx_2 dx_1$ be a 4-form. Determine the integral on the interval $[0, \pi] \times [0, 2\pi] \times [0, 3\pi] \times [0, 4\pi]$.

# CHAPTER 7

# Fundamental Theorem of Calculus

**Carlos Polanco**

Faculty of Sciences, Universidad Nacional Autónoma de México, México

**Abstract** This chapter reviews The Green's, Stokes', and Gauss' Theorems as a direct result of the differentiation and integration operations set out in previous chapters. All the exercises are solved using the Grassmann algebra. The Fundamental theorem of calculus is introduced at the end of this chapter, as an extension of the theorems studied here.

**Keywords:** $dw_1$ form, $dw_2$ form, $w_1$ form, $w_2$ form, divergence, Field associated, Fundamental Theorem of Calculus, Gauss' theorem, Grassmann algebra, Green's theorem, Heaviside-Gibbs algebra, rotational, Stokes' theorem

## 7.1.   Preliminaries

In the following sections, we will introduce the operators and properties of the Grassmann algebra (Chaps. 2-6) with some examples. These operators and their properties are required to introduce the Green's, Stokes', and Gauss' theorems. If the reader is interested in knowing these theorems under the Heaviside Gibbs algebra, he/she can review (Chap. 1).

These theorems derive directly from the integration and differentiation operators of Grassmann algebra and their generalization is provided in the last section of this chapter.

## 7.2.   Green Theorem

**Definition 7.1.** Let $w_1$ be a $1-$form on an open over a region $D \subset \mathbb{R}^2$ bounded by $\partial D$ in the positive perimeter, then

$$\int_{\partial D} w_1 = \int_D dw_1 \qquad (7.1)$$

Green's theorem states that the effect of the vector-valued function $F$ over the oriented closed curve $\partial D$, counter-clockwise orientation (represented by $w_1-$form over $\mathbb{R}^2$), is equivalent to the rotational effect over the area bounded by the region $D$, i.e. $dw_1$.

Proof. The definition of field associated (Sect. 5.3.2) is verified.

Example 7.1. Let $w_1 = -y\,dx + x\,dy$. Verify Green's theorem over the region $c(t) = (\cos t, \sin t), t \in [0\,2\pi]$.

**Solution 7.1.** $\int_{\partial D} w_1 = \int_0^{2\pi} -y\,dx + x\,dy\,dt = \int_0^{2\pi} -\sin t (\cos t)'_t + \cos t (\sin t)'_t\,dt$

$= \int_0^{2\pi} \sin^2 t + \cos^2 t\,dt = 2\pi.$

$$dw_1 = d(-y \wedge dx) + d(x \wedge dy))$$

$$= -\left(\frac{\partial y}{\partial x} dx + \frac{\partial y}{\partial y} dy\right) \wedge dx + \left(\frac{\partial x}{\partial x} dx + \frac{\partial x}{\partial y} dy\right) \wedge dy$$

$$= -\left(\frac{\partial y}{\partial x}\right) dx \wedge dx - \left(\frac{\partial y}{\partial y}\right) dy \wedge dx \qquad (7.2)$$

$$+ \left(\frac{\partial x}{\partial x}\right) dx \wedge dy + \left(\frac{\partial x}{\partial y}\right) dy \wedge dy$$

$$= 2dxdy$$

now, we parameterize $c(r, \theta) = (r\cos\theta, r\sin\theta, 1)$

$$\int_D dw_1 = \int_0^1 \int_0^{2\pi} [2dxdy]\,d\theta dr = \int_0^1 \int_0^{2\pi} \left[2\frac{\partial(x,y)}{\partial(r,\theta)}\right] d\theta dr = \int_0^1 \int_0^{2\pi} 2r\,d\theta dr$$
$$= 2\pi.$$

Note 7.1. $\dfrac{\partial(x,y)}{\partial(r,\theta)} = r.$

The Green's theorem is verified.

## 7.3.  Stokes' Theorem

**Definition 7.2.** Let $w_1$ be a $1-$form on an open over a region $S \subset \mathbb{R}^3$ bounded by $\partial S$ in the positive perimeter, then

$$\int_{\partial S} w_1 = \int_S dw_1 \qquad (7.3)$$

Stokes's theorem states that the effect of the vector-valued function $F$ over the oriented closed curve $\partial D$, counter-clockwise orientation (represented by $w_1-$form over $\mathbb{R}^3$), is equivalent to the rotational effect over the area bounded by the region $D$, i.e. $dw_1$.

Proof. The definition of field associated (Sect. 5.3.2) is verified.

Example 7.2. Let $w_1 = xy\,dx + e^z\,dy + x\,dz$. Verify Stokes' theorem over the region $c(t) = (\cos t, \sin t, 1)$.

**Solution 7.2.** $\displaystyle\int_{\partial D} w_1 = \int_0^{2\pi} xy\,dx + e^z\,dy + x\,dz\,dt = \int_0^{2\pi} \cos t \sin t (\cos t)'_t + e^1$

$\displaystyle(\sin t)'_t + \cos t(1)'_t\,dt = \int_0^{2\pi} -\cos t \sin^2 t + e^t \cos t\,dt = 0.$

$$dw_1 = d(xy \wedge dx) + d(e^z \wedge dy) + d(x \wedge dz)$$

$$= \left(\frac{\partial xy}{\partial x}\,dx + \frac{\partial xy}{\partial y}\,dy + \frac{\partial xy}{\partial z}\,dz\right) \wedge dx$$

$$+ \left(\frac{\partial e^z}{\partial x}\,dx + \frac{\partial e^z}{\partial y}\,dy + \frac{\partial e^z}{\partial z}\,dz\right) \wedge dy$$

$$+ \left(\frac{\partial x}{\partial x}\,dx + \frac{\partial x}{\partial y}\,dy + \frac{\partial x}{\partial z}\,dz\right) \wedge dz$$

$$\text{(7.4)}$$

$$= \left(\frac{\partial xy}{\partial x}\right) dx \wedge dx + \left(\frac{\partial xy}{\partial y}\right) dy \wedge dx + \left(\frac{\partial xy}{\partial z}\right) dz \wedge dx$$

$$+ \left(\frac{\partial e^z}{\partial x}\right) dx \wedge dy + \left(\frac{\partial e^z}{\partial y}\right) dy \wedge dy + \left(\frac{\partial e^z}{\partial z}\right) dz \wedge dy$$

$$= \left(\frac{\partial x}{\partial x}\right) dx \wedge dz + \left(\frac{\partial x}{\partial y}\right) dy \wedge dz + \left(\frac{\partial x}{\partial z}\right) dz \wedge dz$$

$$= -x\,dxdy + dzdx + e^z dzdy$$

now, we parameterize $c(r, \theta) = (r\cos\theta, r\sin\theta, 1)$

$$\int_D dw_1 = \int_0^1 \int_0^{2\pi} [-x\,dxdy + dzdx + e^z dzdy]\,d\theta dr = \int_0^1 \int_0^{2\pi} \left[-r\cos\theta\frac{\partial(x,y)}{\partial(r,\theta)}\right.$$

$$\left. +\frac{\partial(z,x)}{\partial(r,\theta)} + e^1\frac{\partial(z,y)}{\partial(r,\theta)}\right]d\theta dr = \int_0^1 \int_0^{2\pi} -r^2\cos\theta\,d\theta dr = 0.$$

Note 7.2. $\dfrac{\partial(x,y)}{\partial(r,\theta)} = r$, $\dfrac{\partial(z,x)}{\partial(r,\theta)} = 0$, and $\dfrac{\partial(z,y)}{\partial(r,\theta)} = 0$.

The Stokes' theorem is verified.

## 7.4.   Gauss' Theorem

**Definition 7.3.** Let $w_2$ be a $2-$form on an open over a region $\omega \subset \mathbb{R}^3$ bounded by $\partial\omega$, in the positive perimeter, then

$$\iint_{\partial\omega} w_2 = \iiint_\omega dw_2 \qquad\qquad (7.5)$$

Proof. The definition of field associated (Sect. 5.3.3) is verified.

Example 7.3. Let $w_2 = xz\,dxdy - xy\,dxdz - xy\,dydz$. Verify Gauss's theorem over the region $T(\theta,\phi) = (\cos\theta\sin\phi, \sin\theta\sin\phi, \cos\phi)$, $\theta \in [0, 2\pi]$, and $\phi \in [0, \pi]$.

**Solution 7.3.** $\displaystyle\int_D w_2 = \int_0^\pi \int_0^{2\pi} [xz\,dxdy - xy\,dxdz - xy\,dydz]\,d\theta d\phi = \int_0^\pi \int_0^{2\pi} \Big[$

$\displaystyle \cos\theta\sin\phi\cos\phi \frac{\partial(y,x)}{\partial(r,\theta)} + \cos\theta\sin\phi\sin\theta\sin\phi \frac{\partial(z,x)}{\partial(r,\theta)} + \cos\theta\sin\phi \frac{\partial(z,y)}{\partial(r,\theta)} \Big] d\theta d\phi$

$\displaystyle = \int_0^\pi \int_0^{2\pi} d\theta d\phi = -\cos\theta\sin\phi\cos\phi\sin\phi\cos\phi - \cos\theta\sin\phi\sin\theta\sin\phi\sin^2\phi\sin\theta$

$\displaystyle -\cos\theta\sin\phi\sin^2\phi\cos\theta\, d\theta\, d\phi = -\int_0^\pi \int_0^{2\pi} \cos\theta\cos^2\phi\sin^2\phi + \cos\theta\sin^3\phi\sin^2\theta$

$\displaystyle + \cos^2\theta\sin^3\phi\, d\theta\, d\phi = 0.$

Note 7.3. $\dfrac{\partial(x,y)}{\partial(\theta,\phi)} = -\sin\phi\cos\phi$, $\dfrac{\partial(z,x)}{\partial(\theta,\phi)} = -\sin^2\phi\sin\theta$, and $\dfrac{\partial(z,y)}{\partial(\theta,\phi)} = \sin^2\phi\cos\theta$.

$$dw_2 = -d(xy\,dydz) + d(xy\,dzdx) + d(xz\,dxdy)$$

$$= (-\frac{\partial x}{\partial x}dx - \frac{\partial x}{\partial y}dy - \frac{\partial x}{\partial z}dz) \wedge (dy \wedge dz)$$

$$+ (\frac{\partial xy}{\partial x}dx + \frac{\partial xy}{\partial y}dy + \frac{\partial xy}{\partial z}dz) \wedge (dz \wedge dx)$$

$$+ (\frac{\partial xz}{\partial x}dx + \frac{\partial xz}{\partial y}dy + \frac{\partial xz}{\partial z}dz) \wedge (dx \wedge dy)$$

$$= -\frac{\partial x}{\partial x}dx(dy \wedge dz) - \frac{\partial x}{\partial y}dy \wedge (dy \wedge dz) - \frac{\partial x}{\partial z}dz(dy \wedge dz)$$

$$+ \frac{\partial xy}{\partial x}dx(dz \wedge dx) + \frac{\partial xy}{\partial y}dy(dz \wedge dx) + \frac{\partial xy}{\partial z}dz(dz \wedge dx) \tag{7.6}$$

$$+ \frac{\partial xz}{\partial x}dx(dx \wedge dy) + \frac{\partial xz}{\partial y}dy \wedge (dx \wedge dy) + \frac{\partial xz}{\partial z}dz(dx \wedge dy)$$

$$= -\frac{\partial x}{\partial x}dxdydz - \frac{\partial x}{\partial y}dydydz - \frac{\partial x}{\partial z}dzdydz$$

$$+ \frac{\partial xy}{\partial x}dxdzdx + \frac{\partial xy}{\partial y}dydzdx + \frac{\partial xy}{\partial z}dzdzdx$$

$$+ \frac{\partial xz}{\partial x}dxdxdy + \frac{\partial xz}{\partial y}dydxdy + \frac{\partial xz}{\partial z}dzdxdy$$

$$= (2x - 1)\,dxdydz$$

now we parameterize $T(\rho,\theta,\phi) = (\rho\cos\theta\sin\phi, \rho\sin\theta\sin\phi, \rho\cos\phi)$, $\theta \in [0, 2\pi]$, $\phi \in [0, \pi]$, and $\rho \in [0, 1]$.

$$\int_D dw_2 = \int_0^1 \int_0^\pi \int_0^{2\pi} 2x - 1\, dx dy dz\, d\theta\, d\phi\, d\rho = \int_0^1 \int_0^\pi \int_0^{2\pi} (2\rho\cos\theta\sin\phi - 1)$$

$$\frac{\partial(x,y,z)}{\partial(\rho,\theta,\phi)} = \int_0^1 \int_0^\pi \int_0^{2\pi} \rho\cos\phi(2\rho\cos\theta\sin\phi - 1)d\theta\, d\phi\, d\rho = 2\rho^2\cos\phi\sin\phi -$$
$$\rho\cos\phi d\theta\, d\phi\, d\rho = 0.$$

Note 7.4. $\dfrac{\partial(x,y,z)}{\partial(\rho,\theta,\phi)} = \rho\cos\phi$.

The Gauss' theorem is verified.

## 7.5.    Fundamental Theorem of Calculus

**Definition 7.4.** Let $w_k$ be a $k-$form on an open over a region $V \subset \mathbb{R}^n$ bounded by $\partial V$, in the positive area, then

$$\int_{\partial V} w_k = \int_V dw_k \tag{7.7}$$

Proof. The definition of field associated (Sect. 5.3.3) by mathematical induction is verified.

Example 7.4. Let $w_3 = x_3 x_4\, dx_1 dx_2 dx_3$. Verify the Fundamental Theorem of Calculus over the region $T(u_1,u_2,u_3) = (u_1,u_2,u_3,u_1)$, $u_1 \in [0,1]$, $u_2 \in [0,1]$ and $u_3 \in [0,1]$.

**Solution 7.4.**

$$\begin{aligned}
\int_D w_3 &= \int_0^1 \int_0^1 \int_0^1 [x_3 x_4\, dx_1\, dx_2\, dx_3]\, du_1 du_2 du_3 \\
&= \int_0^1 \int_0^1 \int_0^1 \left[ u_1 u_3 \frac{\partial(x_1,x_2,x_3)}{\partial(u_1,u_2,u_3)} \right] du_1 du_2 du_3 \\
&= \int_0^1 \int_0^1 \int_0^1 u_1 u_3 du_1 du_2 du_3 \\
&= \frac{9}{2}
\end{aligned} \tag{7.8}$$

Note 7.5. $\dfrac{\partial(x_1,x_2,x_3)}{\partial(u_1,u_2,u_3)} = 1$.

$$dw_3 = d(x_1 x_4 \, dx_1 dx_2 dx_3 + x_2 x_3 \, dx_3 dx_4 dx_1)$$

$$
\begin{aligned}
&= \left( \frac{\partial x_1 x_4}{\partial x_1} dx_1 + \frac{\partial x_1 x_4}{\partial x_2} dx_2 + \frac{\partial x_1 x_4}{\partial x_3} dx_3 + \frac{\partial x_1 x_4}{\partial x_4} dx_4 \right) \wedge (dx_1 \wedge dx_2 \wedge dx_3) \\
&\quad + \left( \frac{\partial x_2 x_3}{\partial x_1} dx_1 + \frac{\partial x_2 x_3}{\partial x_2} dx_2 + \frac{\partial x_2 x_3}{\partial x_3} dx_3 + \frac{\partial x_2 x_3}{\partial x_4} dx_4 \right) \wedge (dx_3 \wedge dx_4 \wedge dx_1)
\end{aligned}
$$

$$
\begin{aligned}
&= \frac{\partial x_1 x_4}{\partial x_1} dx_1 (dx_1 \wedge dx_2 \wedge dx_3) + \frac{\partial x_1 x_4}{\partial x_2} dx_2 (dx_1 \wedge dx_2 \wedge dx_3) \\
&\quad + \frac{\partial x_1 x_4}{\partial x_3} dx_3 (dx_1 \wedge dx_2 \wedge dx_3) + \frac{\partial x_1 x_4}{\partial x_4} dx_4 (dx_1 \wedge dx_2 \wedge dx_3) \\
&= \frac{\partial x_2 x_3}{\partial x_1} dx_1 (dx_3 \wedge dx_4 \wedge dx_1) + \frac{\partial x_2 x_3}{\partial x_2} dx_2 (dx_3 \wedge dx_4 \wedge dx_1) \\
&\quad + \frac{\partial x_2 x_3}{\partial x_3} dx_3 (dx_3 \wedge dx_4 \wedge dx_1) + \frac{\partial x_2 x_3}{\partial x_4} dx_4 (dx_3 \wedge dx_4 \wedge dx_1)
\end{aligned}
$$

$$
\begin{aligned}
&= \frac{\partial x_1 x_4}{\partial x_1} dx_1 dx_1 dx_2 dx_3 + \frac{\partial x_1 x_4}{\partial x_2} dx_2 dx_1 dx_2 dx_3 \\
&\quad + \frac{\partial x_1 x_4}{\partial x_3} dx_3 dx_1 dx_2 dx_3 + \frac{\partial x_1 x_4}{\partial x_4} dx_4 dx_1 dx_2 dx_3 \\
&\quad + \frac{\partial x_2 x_3}{\partial x_1} dx_1 dx_3 dx_4 dx_1 + \frac{\partial x_2 x_3}{\partial x_2} dx_2 dx_3 dx_4 dx_1 \\
&\quad + \frac{\partial x_2 x_3}{\partial x_3} dx_3 dx_3 dx_4 dx_1 + \frac{\partial x_2 x_3}{\partial x_4} dx_4 dx_3 dx_4 dx_1
\end{aligned}
$$

$$= -(x_1 + x_3) \, dx_1 dx_2 dx_3 dx_4$$

$$(7.9)$$

now, we parameterize $T(u_1, u_2, u_3, u_4) = (u_1, u_2, u_3, u_4), u_1 \in [0, \alpha_1]$, $u_2 \in [0, \alpha_2]$, $u_3 \in [0, \alpha_3]$, and $u_4 \in [0, \alpha_4]$.

$\int_D dw_3 = \int_0^{\alpha_1} \int_0^{\alpha_2} \int_0^{\alpha_3} \int_0^{\alpha_4} (x_1 + x_3) \, dx_1 dx_2 dx_3 dx_4 = \dfrac{1}{4}$. Where $\alpha_1 = \alpha_3 = \alpha_4 = 1$, and $\alpha_2 = \dfrac{1}{2} \left( -1 - \sqrt{3} \right)$, or $\alpha_2 = \dfrac{1}{2} \left( \sqrt{3} - 1 \right)$.

The Fundamental Theorem of Calculus is verified.

## 7.6.   Remarks

The theorems in this chapter are based on the review of **geometric algebra** in $2D$ (Chap. 2), $3D$ space (Chap. 3), and $nD$ space (Chap. 5).

It is important to note that the Fundamental Theorem of Calculus is a generalization of the Green's, Stokes', and Gauss' theorems (Chap. 1), specially the **line integral** and **surface integral**.

We strongly advise the reader reviews and solves all the exercises in this chapter, to get familiar with the **differentiation** operator of this algebra.

In the next chapter, we will review the three applications of exterior calculus.

.

## 7.7.  Exercises

**Exercise 7.1.** Let $w_0 = xyz \, dz \, dy \, dx$ be over the interval $[0,1] \times [0,2] \times [0,3]$. Solve and explain the meaning in Heaviside-Gibbs algebra and Geometric algebra.

**Exercise 7.2.** Let $w_1 = yx \, dx + 2zy \, dy + dz$, evaluate the line integral over the region $c(t) = (\cos t, \sin t), 1), t \in [0, 2\pi]$. Solve with Heaviside-Gibbs algebra and Geometric algebra.

**Exercise 7.3.** Let $w_2 = 2x \, dx \, dy + 3y \, dy \, dz + 4z \, dz \, dx$ evaluate the surface integral over the región $T(r, \theta) = (r \cos \theta, r \sin t, 1), r \in [0,1], \theta \in [0, 2\pi]$. Solve with Heaviside-Gibbs algebra and Geometric algebra.

**Exercise 7.4.** Let $w_1 = -4y \, dx + 4x \, dy$. Verify Green's theorem over the region $c(t) = (\cos t, \sin t), t \in [0, 2\pi]$.

**Exercise 7.5.** Verify Green's theorem over $\int_D xy \, dx + x^2 y^2 \, dy$, where $c$ is the triangle with vertices $(0,0)$, $(2,0)$, and $(0,2)$.

**Exercise 7.6.** Let $w_1 = y \, dx + e^z \, dy + x \, dz$. Verify Stokes' theorem over the region $c(t) = (\cos t, \sin t, 1)$.

**Exercise 7.7.** Verify Stokes' theorem over $\int_D x \, dx + y \, dy + z \, dz$, where $c$ is defined by $c(t) = (\cos t, \sin t, 1 - \cos t - \sin t)$ [1].

**Exercise 7.8.** Let $w_2 = xz \, dx \, dy - xy \, dx \, dz - dy \, dz$. Verify Gauss' theorem over the region $T(\theta, \phi) = (\cos \theta \sin \phi, \sin \theta \sin \phi, \cos \phi)$, $\theta \in [0, 2\pi]$, and $\phi \in [0, \pi]$.

**Exercise 7.9.** Verify Gauss' theorem over $\int_V x^2 \, dy \, dz + y^2 \, dz \, dx + z \, dx \, dy$, where $c$ is defined by $T(\theta, \phi) = (\cos \theta \sin \phi, \sin \theta \sin \phi, \cos \phi)$ [1].

**Exercise 7.10.** Let $w_3 = x_3 x_4 \, dx_1 \, dx_2 \, dx_3$. Verify the Fundamental Theorem of Calculus over the region $T(u_1, u_2, u_3) = (u_1, u_2, u_3, u_1)$, $u_1 \in [0,1]$, $u_2 \in [0,2]$, and $u_3 \in [0,3]$.

# III
# APPLICATIONS

The second part of the book starts with the characterization of the real-valued functions with a review of the concepts of continuity, differentiation, and integration. Integration is presented with mappings on a plane and space. We define the vector-valued functions, their geometric representation and the two vector operators: rotational and divergence. Each section is self-contained so the unfamiliar reader can follow up the subject.

# CHAPTER 8

# Applications

**Carlos Polanco**

Faculty of Sciences, Universidad Nacional Autónoma de México, México

**Abstract** This chapter gives an alternative solution to the spread of an epidemic outbreak of $k$ dimension, using a $k-$Form. The $k-$region, derivative, and integral of this $k-$Form are interpreted. An extension of the $k$ dimension is proposed using a $k-$Form equivalent to the electric current and the magnetic field, known as Ampere's law. An algorithm to determine the main function of a protein is introduced using a $k-$Form. Finally, the $k-$region, derivative, and integral of this $k-$Form are interpreted.

**Keywords**: Ampere's law, clinical variables, mathematical epidemiology, non-clinical variables, structural proteomics

## 8.1. Mathematical Epidemiology

## 8.1.1. Preliminaries

In recent years, after the emblematic analysis of 335 infectious emerging diseases from 1940 to 2004, in which it was reported that 60% were zoonosis and 25% were mosquito-borne viruses [30], and after the A-H1N1 flu outbreak of 1989 [31], there has been substantial progress in the development of surveillance systems of serious diseases with epidemic potential to support public health, clinical infrastructure, and the limited responsiveness of Emergency Services.

At present, it is still uncertain if a sporadic zoonosis restricted to a certain area will become a global pandemic or something in between. Therefore, surveillance systems of severe infectious diseases with epidemic potential should not only be based on the number of notified cases and their space-time distribution in a determined geographical area, to issue an early warning.

The best would be to also consider non-clinical variables, such as socio-demographic factors, public transport, livestock production, and vaccinated population, as it is known [31] that these factors are the epidemiological foundation for the spread of a potential pandemic outbreak. Today, a person can be infected on one continent and be on another 10 hours later.

This, combined with the virulence of the pathogenic agents and some socio-demographic factors, determine their spreading capability. A surveillance system of severe infection diseases with epidemic potential, will give health authorities valuable time to promote suitable measures and minimize the spread of the disease.

For this reason, it is expected that a surveillance system of severe infectious diseases with epidemic potential identifies, as soon as possible, specific symptomatic cases of an infectious process; this requires a predictive element that foresees, with a certain degree of accuracy, a possible event in the time/space of this infectious process so the authorities take preventive measures in the affected region.

In our opinion, two of the main factors undermining the effectiveness of the warnings are, on the one hand, the increasingly efficient means of transport and on the other, the numerous mild diseases e.g. colds that present fever.

Nowadays, the surveillance systems of serious infectious diseases with epidemic potential are mainly based on the number of microbiologically [32] verified cases; the warnings, although real, are also late as monitoring is based on the assumption that symptomatic subjects will go to a clinic.

However, if the transmissibility and/or lethality of the virus is very high, or if the number of medical facilities in the area is very limited, the index patient and some of his/her contacts will probably die before receiving medical attention, which will make even harder to trace back the contacts net that will continue growing. Additionally, the number of doctors and clinics available is frequently less than optimal, as in the case of developing countries, where the population does not usually seek medical advice for many different reasons.

In this circumstance, it is necessary to have a predictive model of serious infectious diseases with pandemic potential that considers and weights clinical and non-clinical variables, instead of depending only on the number of microbiologically confirmed cases, and that forecasts the emergence and progress of the outbreak in a region.

## 8.1.2. Model

The model proposed, defines a $k-$form function (Eq. 5.20), whose $dx_k$ is a measure of the net flux through the boundary of an infinitesimal $(k+1)-$region, enclosed in an oriented $k-$geographical region that represents the effect of the total flux on a particular area of that $k-$geographical region, where

**Definition 8.1.** The derivative [3] of a 3$-$form function $w_k$ (Eq. 8.1) is a $k+1-$form function of $C^1$ class $w_{k+1} = dw_k$ (Eq. 8.2).

$$w_k = f(x_1, \cdots, x_n) \, dx_j \wedge \cdots \wedge dx_k \qquad (8.1)$$

$$dw_k = \left( \sum_{i=1}^{n} \frac{\partial f(x_1, \cdots, x_n)}{\partial x_i} \, dx_i \right) \wedge dx_j \wedge \cdots \wedge dx_k \qquad (8.2)$$

and,

**Definition 8.2.** The integral of a differential form $w_k$ in $\mathbb{R}^n$ over a region $D \in \mathbb{R}^n$, is represented by (Eq. 8.3).

$$\int_D w_k = \int_{a_1}^{a_k} \cdots \int_{a_{k+m}}^{a_n} f_1(x_1, \cdots, x_n) \, dx_1 \cdots dx_2 + \cdots +$$
$$f_n(x_1, \cdots, x_n) \, dx_n \cdots dx_1 \, dD \quad (8.3)$$

The previous definitions are in (Chaps. 5, and 6) if the reader wants to deepen in these concepts, it is advisable to review these chapters.

### 8.1.2.1   Clinical Variables

Clinical variables [31] are parameters strongly associated with an epidemic process and they are related to the seriousness of the patients' condition, or the medical supplies necessary for their attention, i.e. hemodynamic monitors and mechanical ventilators.

### 8.1.2.2   Non-Clinical Variables

Non-clinical variables associated with an epidemic process are those variables that are not associated with the medical aspect and may well be associated with transport phenomena, education, population growth, or accessibility to drinking water, e.g. passengers traveling, illiterate indigenous population, immigrant population, and dwellings without piped water.

## 8.1.3.   Algorithm

The function is a vector-valued function $f : \mathbb{R}^k \to \mathbb{R}^k$, where $k$ is the number of clinical and non-clinical variables.

The integral 6.6 of a $k-$form represents the total effect or flux that a vector-valued function $f(x)$ has over the oriented $k-$volume, on an interval of the domain of the function; and the derivative (Def. 5.3.5) $dx_k$ is a measure of the net flux through the boundary of an infinitesimal $(k+1)-$region enclosed in an oriented $k-$volume.

### 8.1.4.  Discussion

A method that estimates an epidemic outbreak is a multifactorial phenomenon, its dissemination can certainly be measured by counting the number of infected people and the deaths, however, this approach is very expensive. In the case of viral phenomena, particularly of the coronavirus type, the infection rate can grow very fast and become uncontrollable.

In this sense, algorithms oriented to estimate early warnings with epidemic (or pandemic) potential, should cover multiple variables of very different nature. This is only possible in orientable regions located in $k$ dimension spaces that are ideal for the re-creation of these phenomena.

## 8.2.  Structural Proteomics

### 8.2.1.  Preliminaries

Structural proteomics is a discipline that focuses on the identification and classification of proteins, and although it started with experimental techniques in laboratories, today it is done with mathematical-computational programs that evaluate, with different algorithms, the proteins and their amino acids or nucleotides (genes), as in this work.

This type of algorithm acts in the three-dimensional expression where the protein acts, or in its linear representation called sequence.

After this classification, and in order of importance, the algorithms are classified in supervised or non-supervised. A supervised algorithm is one whose metric requires some form of human assistance, while those that do not require any assistance are called non-supervised algorithms.

Here is an approximation to a non-supervised algorithm, whose metric uses a differential form $w$.

### 8.2.2.  Model

The model described below, defines a $k-$form function (Eq. 5.20), whose $dx_k$ is a measure of the net flux through the boundary of an infinitesimal $(k+1)-$ region enclosed in an oriented $k-$ geographical region that represents the effect of the total flux over a particular area of that $k-$geographical region.

**Definition 8.3.** The derivative [3] of a $k-$form function $w_k$ (Eq. 8.4) is a $k+1-$form function of $C^1$ class $w_{k+1} = dw_k$ (Eq. 8.5), where

$$w_k = f(x_1, \cdots, x_n)\, dx_j \wedge \cdots \wedge dx_k \tag{8.4}$$

$$dw_k = \left( \sum_{i=1}^{n} \frac{\partial f(x_1, \cdots, x_n)}{\partial x_i} \, dx_i \right) \wedge dx_j \wedge \cdots \wedge dx_k \qquad (8.5)$$

and

**Definition 8.4.** The integral of a differential form $w_k$ in $\mathbb{R}^n$ over a region $D \in \mathbb{R}^n$, is represented by Eq. 8.6.

$$\int_D w_k = \int_{a_1}^{a_k} \cdots \int_{a_{k+m}}^{a_n} f_1(x_1, \cdots, x_n) \, dx_1 \cdots dx_2 + \cdots +$$
$$f_n(x_1, \cdots, x_n) \, dx_n \cdots dx_1 \, dD \quad (8.6)$$

The previous definitions correspond to Chaps. 5, and 6, if the reader wants to deepen in these concepts it is advisable to review these chapters.

## 8.2.3.   Algorithm

A function is a vector-valued function $f : \mathbb{R}^k \to \mathbb{R}^k$, where $k$ is the number of polar interactions.

The integral 6.6 of a $k-$form represents the total effect or the flux that a vector-valued function $f(x)$ has over the oriented $k-$volume, on an interval of the domain of the function; and the derivative (Def. 5.3.5) $dx_k$ is a measure of the net flux through the boundary of an infinitesimal $(k+1)-$region enclosed in an oriented $k-$volume.

## 8.2.4.   Discussion

A method that estimates the predominant function of a protein is a multifactorial phenomenon. This can be obtained by studying a fundamental physico-chemical property, such as the polarity/charge of the amino acids in their sequence, thoroughly evaluating the protein and not only by obtaining a numerical value.

In this sense, algorithms oriented to estimate the predominant function of a protein should be a matrix or vector, so all possibilities are evaluated. With this in mind, it will be necessary to define the metric in oriented regions of $k$ dimension. The algebra studied in this book will be applied to this purpose.

## 8.3.　Ampere's Law

### 8.3.1.　Preliminaries

The physicist and mathematician Andre-Marie Ampere (1775–1836) stated one of the main theorems of electromagnetism that is often considered as the magnetic equivalent of Gauss' theorem.

He explained that static electricity and magnetism have a common origin, but different effect. The common origin is an electric charge or steady charge, when these charges are in motion they induce a magnetic field. He also observed that there are materials that exhibit continuous movement of these charges and are, therefore, permanent magnets. Thus, Andre-Marie Ampere created a new field named Electrodynamics.

Ampere's law relates a static magnetic field to the cause producing it, i.e. a steady electric current.

$$\oint_C \vec{H} \cdot d\vec{l} = \iint_S \vec{J} \cdot d\vec{S} + \frac{d}{dt} \iint_S \vec{D} \cdot d\vec{S} \tag{8.7}$$

Where the last term is the displacement current, provided that this current is constant and directly proportional to the magnetic field and the integral is (E) multiplied by its relative mass [33].

### 8.3.2.　Model

The model proposed below defines a $k-$form function (Eq. 5.20), whose $dx_k$ is the measure of the net flux through the boundary of an infinitesimal $(k+1)-$ region, enclosed in an oriented $k-$geographical region that represents the effect of this total flux over a particular area of that $k-$geographical region,
where

**Definition 8.5.** The derivative [3] of a $k-$form function $w_k$ (Eq. 8.8) is a $k-1-$form function of $C^1$ class $w_{k+1} = dw_k$ (Eq. 8.9).

$$w_k = f(x_1, \cdots, x_n)\, dx_j \wedge \cdots \wedge dx_k \tag{8.8}$$

$$dw_k = \left( \sum_{i=1}^{n} \frac{\partial f(x_1, \cdots, x_n)}{\partial x_i}\, dx_i \right) \wedge dx_j \wedge \cdots \wedge dx_k \tag{8.9}$$

and,

**Definition 8.6.** The integral of a differential form $w_k$ in $\mathbb{R}^n$ over a region $D \in \mathbb{R}^n$, is represented by Eq. 8.10.

$$\int_D w_k = \int_{a_1}^{a_k} \cdots \int_{a_{k+m}}^{a_n} f_1(x_1, \cdots, x_n)\, dx_1 \cdots dx_2 + \cdots +$$

$$f_n(x_1, \cdots, x_n)\, dx_n \cdots dx_1\, dD \quad (8.10)$$

These definitions are found in (Chaps. 5, and 6) if the reader wants to deepen in this concept, it is advisable to review these chapters.

Ampere's work arose a question, can a magnetic field generate an electric current? Michael Faraday (1791-1867) proved it can.

With these two contributions, it was possible to know the amount of electric current that we can get from the variation of a magnetic flow. This made possible the large-scale generation of electrical energy.

### 8.3.3.   Algorithm

The function is a vector-valued function $f : \mathbb{R}^k \to \mathbb{R}^k$, where $k$ is the number of clinical and non-clinical variables.

The integral 6.6 of a $k-$form represents the total effect or flux that a vector-valued function $f(x)$ has over an oriented $k-$volume on an interval of the domain of the function, and the derivative (Def. 5.3.5) $dx_k$ is a measure of the net flux through the boundary of an infinitesimal $(k+1)-$region enclosed in an oriented $k-$volume.

### 8.3.4.   Discussion

This approach extends to a $k$ dimension the oriented region of ampere's law. Although its use solves cases in three-dimensional space or plane, its generalization requires the algebra studied in this book if the oriented region is in this dimension.

## 8.4.   Remarks

The reader will find that although the solution to any of the problems found here is extensive, it is theoretically possible.

Therefore, these three applications cannot be solved with **Heaviside-Gibbs algebra** as it is impossible to orient the regions to higher spaces, however, the solution is possible if we use **Geometric algebra** (or **Grassmann algebra** [2, 9, 20, 22, 23, 24, 26].

In these chapters (Chaps. 5-7), we showed how the **Geometric algebra** or **Grassmann algebra** is used in the generalization of the $n$ dimensional space.

# CHAPTER 9

# SOLUTIONS

## Solutions Chapter 1

**Solution 1.1.** (i) The map is $T : \mathbb{R} \Rightarrow \mathbb{R}^2, (a\cos\theta, b\sin\theta)$, $\theta \in [0, 4\pi]$. (ii) See (Fig. 1.2).

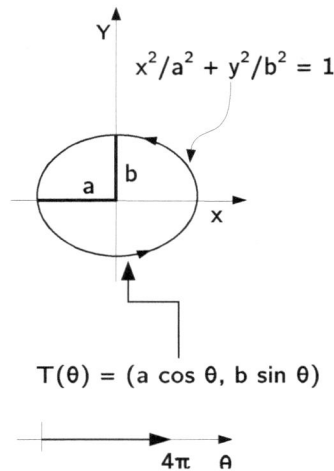

**Figure 1.2** Map of the ellipse where $b < a$. Figure adapted from [1].

(iii) The mapping runs twice the perimeter of the ellipse.

**Solution 1.2.** The map is $T : \mathbb{R} \Rightarrow \mathbb{R}^3, (\cos\theta, \sin\theta, 1 - \cos\theta - \sin^3\theta)$, $\theta \in [0, 3\pi]$.

**Solution 1.3.** (i) The paraboloid is the graph of the function $f(x,y) = 1 - x^2 - y^2$, the map $T : \mathbb{R}^2 \Rightarrow \mathbb{R}^2, (r\cos\theta, r\sin\theta)$, $r \in [0,1], \theta \in [0, 2\pi]$ transforms the rectangle into the unit circle and $f \circ T$ is the third component of the map $T : \mathbb{R}^2 \Rightarrow \mathbb{R}^3$. Then, $T(r, \theta) = (r\cos\theta, r\sin\theta, 1 - r^2\cos^2\theta - r^2\sin^2\theta) = (r\cos\theta, r\sin\theta, 1 - r^2)$.

**Solution 1.4.** (i) $\oint_C F \circ c(t) \cdot c'(t) \, dt = \int_0^{2\pi} (-\sin^2 t, \cos t) \cdot (-\sin t, \cos t) \, dt = \pi.$

**Solution 1.5.** (i) $\oint_C F \circ c(t) \cdot c'(t)\,dt = \int_0^{2\pi}(-\sin^2 t, \cos t, 1)\cdot(-\sin t, \cos t, 0)\,dt = \pi$.

**Solution 1.6.** (i) $T(r,\theta) = (r\cos\theta, r\sin\theta, \sqrt{1-r^2})$. $\oiint_S F \circ T(u,v)\cdot\eta(u,v)\,dS = \int_0^{2\pi}\int_0^1 (r\sin\theta, r\cos\theta, \sqrt{1-r^2})\cdot(\frac{-r^2\cos\theta}{\sqrt{1-r^2}}, \frac{r^2\sin\theta}{\sqrt{1-r^2}}, r)\,dr\,d\theta = \frac{2}{3}\pi$. (ii) $T(t) = (\cos t, \sin t, 0)$ $\oint_D F \circ T(t)\cdot T'(t)\,dt = \int_0^{2\pi}(\sin t, \cos t, 0)\cdot(-\sin t, \cos t, 0)\,dt = 0$.

**Solution 1.7.** $\int_0^1 \int_0^{2\pi} \sqrt{(\cos\theta, \sin\theta, 0)\times(-r\sin\theta, r\cos\theta, 0)}\,d\theta\,dr = \pi$.

**Solution 1.8.** From Green' theorem

$$\oint_{\partial C} F \circ c(t)\cdot c'(t)\,dt = \iint_C (\nabla\times F)\cdot\mathbf{k}\,dy\,dx.$$

Then $F(x,y) = (2xy - x^2, x + y^2)$ and the first mapping is $T_1(t) = (t, t^2), t\in[0,1]$ so $\int_0^1(2t^3 - t^2, t + t^4)\cdot(1, 2t)\,dt = \frac{7}{6}$. The second mapping is $T_2(t) = (t, \sqrt{t}), t\in[1,0]$, then $\int_1^0(2t^{\frac{3}{2}} - t^2, 2t)\cdot(1, \frac{1}{2\sqrt{t}})\,dt = -\frac{17}{15}$. So the line integral is $\frac{1}{30}$. The double integral is $\iint_C(\nabla\times F)\cdot\mathbf{k}\,dy\,dx = \int_0^1\int_{x^2}^{\sqrt{x}}(0, 0, 1 - 2x)\cdot\mathbf{k}\,dy\,dx = \frac{1}{30}$. The double integral is equal to the line integral, the Green' theorem is verified.

**Solution 1.9.** From Stokes' theorem

$$\oint_{\partial D} F \circ c(t)\cdot c'(t)\,ds = \iint_S (\nabla\times F)(T_\beta)\cdot T_v\times T_u\,dv\,du.$$

Since $\frac{x^2}{2} + \frac{y^2}{2} = 2 \Leftrightarrow x^2 + y^2 = 4$, the mapping is $T(t) = (2\cos t, 2\sin t, 2), t\in[2\pi, 0]$. Then $-\int_0^{2\pi}(6\sin t, -4\cos t, 8\sin t)\cdot(-2\sin t, 2\cos t, 0)\,dt = \int_0^{2\pi} -12\sin^2 t - 8\cos^2 t\,dt = 20\pi$. With the mapping $T_\beta(r,\theta) = (r\cos\theta, r\sin\theta, \frac{r^2}{2})$, the double integral is $\iint_C(\nabla\times F)(T_\beta)\cdot T_{\beta r}\times T_{\beta\theta}\,dr\,d\theta = \int_0^2\int_0^{2\pi}(2r\sin\theta + r\cos\theta, 0, -\frac{r^2}{2} - 3)\cdot(r^2\cos\theta, r^2\sin\theta, -r)\,d\theta\,dr = 2\pi$. The double integral is equal to the line integral, the Stokes' theorem is verified.

**Solution 1.10.** From Gauss' theorem

$$\oiint_{\partial W} F \circ T(u,v)\cdot T_v\times T_u\,dv\,du = \iiint_W (\nabla\cdot F)\,dz\,dy\,dx.$$

First we compute $\nabla\cdot F = 2xz^3 + 2xz^3 + 4xz^3 = 8xz^3$, then

$$\begin{aligned}
\oiint_S F\,ds &= \iiint_B (\nabla\cdot F)\,dV \qquad\qquad (1.1)\\
&= \int_{-3}^3\int_{-2}^2\int_{-1}^1 8xz^3\,dx\,dy\,dz\\
&= 0.
\end{aligned}$$

# Solutions for Chapter 2

**Solution 2.1.** $v = -\sigma_2 + \sigma_1 \wedge \sigma_2$, $v = \sigma_2 + 2\sigma_{21}$, $v = 1 + \sigma_1 + 2\sigma_2 - \sigma_{12}$.

**Solution 2.2.** $a = -\sigma_1 + \sigma_2$, and $b = 2\sigma_1 + 3\sigma_2$.

**Solution 2.3.** $ab = -\sigma_2 + \sigma_1 = \sigma_1 - \sigma_2$. So $a \wedge b = \sigma_1 - \sigma_2$.

**Solution 2.4.** $ab = 1 - 3\sigma_{12}$, and $ba = 1 + 3\sigma_{12}$, $a \wedge b = \dfrac{1}{2}(ab - ba) = -3\sigma_{12}$.

**Solution 2.5.** (i) $ab = \sigma_2, ba = -\sigma_2$. (ii) $a \cdot b = 0$. (iii) $a \wedge b = \sigma_2$.

**Solution 2.6.** (i) $a(b+c) = 2$. (ii) $ab = 1 - \sigma_{12}$ and $ac = 1 + \sigma_{12}$, then $ab + ac = 2$. (iii) From (i) and (ii) yes, it is.

**Solution 2.7.** (i) $a(b+c) = 1 - \sigma_1 + \sigma_2 + 3\sigma_{12}$. $(b+c)a = 1 + \sigma_1 + \sigma_2 + 3\sigma_{12}$. (ii) $a \wedge (b+c) = \sigma_1 - \sigma_2$. (iii) $a \wedge b = -\sigma_2$ and $a \wedge c = 0$, then $a \wedge b + a \wedge c = \sigma_1 - \sigma_2$. (iv) Yes, it is.

**Solution 2.8.** $a^{-1} = \dfrac{a}{a \cdot a} = \dfrac{2 - \sigma_1 + \sigma_1 \sigma_2}{2 - 4\sigma_1 + 4\sigma_{12}}$.

**Solution 2.9.** (i) From the definition, the reversion of $a$ is $a^{\dagger} = \sigma_2 \sigma_1$.

**Solution 2.10.** Its blades are $< a >_0 = 1$ and $< a >_1 = 0 < a >_2 = 2\sigma_{12}$.

**Solution 2.11.** $a \wedge b = -2\sigma_{12} + \sigma_1$ then $I(a \wedge b) = \sigma_{12}(-2\sigma_2 + \sigma_1) = -2\sigma_1 - \sigma_2$.

**Solution 2.12.**

$$
\begin{aligned}
||a|| &= \sqrt{\lfloor aa^{\dagger} \rfloor} \\
&= \sqrt{\lfloor (1 + 2\sigma_1 + 3\sigma_2 + 3\sigma_{21})(1 + 2\sigma_1 + 3\sigma_2 - 3\sigma_{21}) \rfloor} \\
&= \sqrt{\lfloor 5 + \cancel{22\sigma_1 - 6\sigma_2} \rfloor} \\
&= \sqrt{5}.
\end{aligned}
\tag{2.2}
$$

**Solution 2.13.** (i) $a(bc) = -\alpha\sigma_2 - \sigma_1$. (ii) $(ab)c = -\alpha\sigma_2 - \sigma_1$. (iii) From the results (i) and (ii) yes, it is.

**Solution 2.14.** (i) $Ia = \sigma_1\sigma_2 a = \sigma_2 + \sigma_1$. (ii) $aI = a\sigma_1\sigma_2 = -\sigma_2 - \sigma_1$. (iii) From these results, (i) is a **rotation** of $\frac{\pi}{2}$ in the **clockwise** direction and (ii) is a **rotation** of $\frac{\pi}{2}$ in the **counter-clockwise** direction.

**Solution 2.15.** (i) $IIa = \sigma_1\sigma_2\sigma_1\sigma_2 a = -2\sigma_1 - 3\sigma_2$. (ii) $aII = a\sigma_1\sigma_2\sigma_1\sigma_2 = -2\sigma_1 - 3\sigma_2$. (iii) From these results, (i) is a **reflection** of $\pi$ in the **clockwise** direction and (ii) is a **reflection** of $\pi$ in the **counter-clockwise** direction.

**Solution 2.16.** $u = \dfrac{u}{||u||} = \dfrac{\sigma_1 - 2\sigma_2}{3}$. Then $y = -uxu = \dfrac{10}{9}\sigma_1 + \dfrac{5}{9}\sigma_2$.

**Solution 2.17.** The **line** $L_{x_0}(v)$ is given by

$$L_{x_0=(1,1)}(v) := \{\mathbf{x} \mid (\mathbf{x} - x_0) \wedge v = 0\}$$

$$[(x_1\sigma_1 + x_2\sigma_2) - (\sigma_1 + \sigma_2)] \wedge \sigma_1 = 0$$
$$[(x_1 - 1)\sigma_1 + (x_2 - 1)\sigma_2] \wedge \sigma_1 = 0$$

The exterior product $(x - x_0) \wedge v = \frac{1}{2}[(x - x_0)v - v(x - x_0)]$,

$$[(x_1 - 1)\sigma_1 + (x_2 - 1)\sigma_2] \wedge \sigma_1 = (x_2 - 1)\sigma_1\sigma_2 = 0 \tag{2.3}$$

From (Eq. 2.3), $x_1 = \mathbb{R}$ and $x_2 = 1$, so the points with the form $(\mathbb{R}, 1)$ are the solution. Note that the point $(1, 1)$ meets the **line** $L_{x_0}(v)$.

**Solution 2.18.** The **plane** $P_{x_0}(u, v)$ is given by

$$P_{x_0=(2,1)}(u \wedge v) := \{\mathbf{x} \mid (\mathbf{x} - x_0) \wedge (u \wedge v) = 0\} \tag{2.4}$$

$$[(x_1\sigma_1 + x_2\sigma_2) - (\sigma_1 + 2\sigma_2)] \wedge (\sigma_1\sigma_2) = 0$$
$$[(x_1 - 1)\sigma_1 + (x_2 - 2)\sigma_2] \wedge (\sigma_1\sigma_2) = 0 \tag{2.5}$$

From the equation $[(x_1 - 1)\sigma_1 + (x_2 - 2)\sigma_2][\sigma_1\sigma_2] - [\sigma_1\sigma_2][(x_1 - 1)\sigma_1 + (x_2 - 2)\sigma_2] = 0$. So, the points $(x_2 - 2, x_1 - 1)$ are the solution. Note that the point $(2, 1)$ meets the **plane** $P_{x_0=(2,1)}(u \wedge v)$.

# Solutions for Chapter 3

**Solution 3.1.** $v = 2\sigma_{123}$, $v = 1 + 3\sigma_{321}$, $v = 1 + \sigma_1 + 2\sigma_2 - \sigma_{12} + \sigma_{123}$.

**Solution 3.2.** $a = -\sigma_1 + \sigma_2 + \sigma_3$, and $b = 2\sigma_1 + 3\sigma_2 - 3\sigma_3$.

**Solution 3.3.** $ab = \sigma_{23} - \sigma_{13}$ $ba = \sigma_{23} - \sigma_{23}$. So $a \wedge b = 0$.

**Solution 3.4.** $ab = -\sigma_{3213} = -\sigma_{21} = \sigma_{12}$, and $ba = -\sigma_{12}$, $a \wedge b = \dfrac{1}{2}(ab - ba) = -\sigma_{12}$.

**Solution 3.5.** (i) $ab = \sigma_{21}, ba = -\sigma_{21}$. (ii) $a \cdot b = 0$. (iii) $a \wedge b = \sigma_{21} = -\sigma_{12}$.

**Solution 3.6.** (i) $a(b+c) = 2\sigma_{31} - \sigma_{32} + \sigma_{12}$. (ii) $ab = \sigma_{31} - \sigma_{32}$ and $ac = \sigma_{31} + \sigma_{12}$, then $ab + ac = 2\sigma_{31} - \sigma_{32} + \sigma_{12}$. (iii) From (i) and (ii) yes, it is.

**Solution 3.7.** (i) $a(b+c) = 1 - \sigma_1 + \sigma_2 + 3\sigma_{12}$. $(b+c)a = 1 + \sigma_1 + \sigma2 + 3\sigma_{12}$. (ii) $a \wedge (b+c) = \sigma_1 - \sigma_2$. (iii) $a \wedge b = -\sigma_2$ and $a \wedge c = 0$, then $a \wedge b + a \wedge c = \sigma_1 - \sigma_2$. (iv) Yes, it is.

**Solution 3.8.** $a^{-1} = \dfrac{a}{a \cdot a} = \dfrac{1 - \sigma_1 + 2\sigma_1\sigma_3}{-2 + 4\sigma_{12} - 4\sigma_3}$.

**Solution 3.9.** (i) From the definition, the reversion of $a$ is $a^\dagger = -\sigma_1\sigma_3$.

**Solution 3.10.** Its blades are $< a >_0 = 1$ and $< a >_2 = 2\sigma_{12}$ $< a >_3 = -\sigma_{123}$.

**Solution 3.11.** $a \wedge b = -2\sigma_3$ then $I(a \wedge b) = \sigma_{123}(-2\sigma_3) = -2\sigma_{12}$.

**Solution 3.12.** The norm of $a$ is.

$$
\begin{aligned}
||a|| &= \sqrt{\lfloor aa^\dagger \rfloor} \\
&= \sqrt{\lfloor (1 + \sigma_1 + \sigma_2 - \sigma_{21})(1 + \sigma_1 + \sigma_2 + \sigma_{21}) \rfloor} \\
&= \sqrt{\lfloor 4 + \cancel{4\sigma_2} \rfloor} \\
&= \sqrt{2}.
\end{aligned}
\tag{3.6}
$$

**Solution 3.13.** (i) $a(bc) = -\alpha\sigma_{32} - \sigma_{31}$. (ii) $(ab)c = -\alpha\sigma_{32} - \sigma_{31}$. (iii) From the results (i) and (ii) yes, it is.

**Solution 3.14.** (i) $Ia = \sigma_1\sigma_2\sigma_3 a = \sigma_{23} + \sigma_{12}$. (ii) $aI = a\sigma_1\sigma_2\sigma_3 = \sigma_{23} + \sigma_{12}$. (iii) From these results, (i) is a **rotation** of $\frac{\pi}{2}$ in the **clockwise** direction and (ii) is a **rotation** of $\frac{\pi}{2}$ in the **counter-clockwise** direction.

**Solution 3.15.** (i) $IIa = \sigma_{1231232} + \sigma_{1231233} = -\sigma_1 - \sigma_3$. (ii) $aII = \sigma_{2123123} + \sigma_{3123123} = -\sigma_1 - \sigma_3$. (iii) From these results, (i) is a **reflection** of $\pi$ in the **clockwise** direction and (ii) is a **reflection** of $\pi$ in the **counter-clockwise** direction.

**Solution 3.16.** $u = \dfrac{u}{||u||} = \dfrac{\sigma_1 - 2\sigma_2}{3}$. Then $y = -uxu = \dfrac{2}{3}\sigma_1 + \dfrac{8}{9}\sigma_2 + \sigma_3$.

**Solution 3.17.** The **line** $L_{x_0}(v)$ is given by

$$L_{x_0=(0,1,0)}(v) := \{\mathbf{x} \mid (\mathbf{x} - x_0) \wedge v = 0\}$$

$$[(x_1\sigma_1 + x_2\sigma_2 + x_3\sigma_3) - (0\sigma_1 + \sigma_2 + 0\sigma_3)] \wedge (\sigma_1 + \sigma_2 + \sigma_3) = 0$$
$$[(x_1 - 0)\sigma_1 + (x_2 + 1)\sigma_2 + (x_3 - 0)\sigma_3] \wedge (\sigma_1 + \sigma_2 + \sigma_3) = 0$$

The outer product $(x - x_0) \wedge v = \frac{1}{2}[(x - x_0)v - v(x - x_0)]$,

$$(x_2 - 1)\sigma_{23} + x_1\sigma_{13} = 0$$

$$(3.7)$$

From (Eq. 4.11), $x_1 = 0, x_2 = 1$, and $x_3 = \mathbb{R}$. So, the points with the form $(0, 1, \mathbb{R})$ are the solution. Note that the point $(0, 1, 0)$ meets the **line** $L_{x_0}(v)$.

**Solution 3.18.** The **plane** $P_{x_0}(u, v)$ is given by

$$P_{x_0=(2,1,1)}(u \wedge v) := \{\mathbf{x} \mid (\mathbf{x} - x_0) \wedge (u \wedge v) = 0\} \qquad (3.8)$$

$$[(x_1\sigma_1 + x_2\sigma_2) - (\sigma_1 + 2\sigma_2)] \wedge (\sigma_1\sigma_2) = 0$$
$$[(x_1 - 1)\sigma_1 + (x_2 - 2)\sigma_2] \wedge (\sigma_1\sigma_2) = 0 \qquad (3.9)$$

From (Eq. 4.13), $[(x_1 - 1)\sigma_1 + (x_2 - 2)\sigma_2][\sigma_1\sigma_2] - [\sigma_1\sigma_2][(x_1 - 1)\sigma_1 + (x_2 - 2)\sigma_2] = 0$. So, the points $(x_2 - 2, x_1 - 1)$ are the solution. Note that the point $(2, 1, 1)$ meets the **plane** $P_{x_0=(2,1,1)}(u \wedge v)$.

## Solutions for Chapter 4

**Solution 4.1.** $v = 2$, $v = 1 + 3\sigma_1$, $v = -\sigma_1 \wedge \sigma_2 \wedge \sigma_3 \wedge \sigma_4 \wedge \sigma_5 \wedge \sigma_6$, $v = 2\sigma_{21}$, $v = 1 + \sigma_1 + 2\sigma_2 - \sigma_{12} + \sigma_{123456789}$.

Remark 4.1. All of the above vectors are considered to be **multivectors** in their most general sense. So it is avoided to qualify them particularly as **bivectors**, **trivectors** among other adjectives.

**Solution 4.2.** (i) $a = -\sigma_1 + \sigma_2 + \sigma_3 - \sigma_4 + 2\sigma_5$, and $b = 2\sigma_1 + 3\sigma_2 - 3\sigma_3 + \sigma_4 - \sigma_5 + 3\sigma_6$.

(ii) The line generated by $L_{x_0}$ and the plane generated by $P_{x_0}$, with the orientation of the vectors $v$ and $u \wedge v$ respectively.

$$L_{x_0 = (1,2,3,4)}(v) = \{\mathbf{x} \mid (\mathbf{x} - x_0) \wedge v = (4,3,2,1) = 0\},$$

$$P_{x_0 = (1,2,3,4)}(u \wedge v) = \{\mathbf{x} \mid (\mathbf{x} - x_0) \wedge [u = (1,-1,1,-1) \wedge v = (1,2,-2,3)] = 0\}.$$

**Solution 4.3.** $ab = \sigma_{2345}$, and $ba = \sigma_{2345}$, $a \wedge b = \dfrac{1}{2}(ab - ba) = 0$.

**Solution 4.4.** $ab = \sigma_{2345}$, and $ba = \sigma_{2345}$, $a \cdot b = \dfrac{1}{2}(ab + ba) = \sigma_{2345}$.

**Solution 4.5.** (i) $ab = \sigma_{12345678}, ba = -\sigma_{56781234} = \sigma_{12345678}$. (ii) $a \cdot b = \sigma_{12345678}$. (iii) $a \wedge b = 0$.

**Solution 4.6.** (i) $a(b + c) = \sigma_4 - 2\sigma_{234} - \sigma_{134}$. (ii) $ab = -\sigma_{234} - \sigma_{134}$ and $ac = -\sigma_{234} + \sigma_4$, then $ab + ac = \sigma_4 - 2\sigma_{234} - \sigma_{134}$. (iii) From (i) and (ii) yes, it is.

**Solution 4.7.** (i) $a(b + c) = -\sigma_{234} - \sigma_{34}$. $(b + c)a = \sigma_{234} - \sigma_{34}$. (ii) $a \wedge (b + c) = -\sigma_{234}$. (iii) $a \wedge b = -\sigma_{234}$ and $a \wedge c = -\sigma_{34}$, then $a \wedge b + a \wedge c = -\sigma_{234} - \sigma_{34}$. (iv) Yes, it is.

**Solution 4.8.** $a^{-1} = \dfrac{a}{a \cdot a} = \dfrac{1 + \sigma_1 + \sigma_1 + \cdots + \sigma_n}{a + \sigma_1 a + \sigma_2 a + \cdots + \sigma_n a} = \dfrac{a}{a + a(\sigma_1 + \sigma_2 + \cdots + \sigma_n)}$
$= \dfrac{a}{a + a(a - 1)} = \dfrac{a}{a + a^2 - a)} = \dfrac{1}{a}$.

**Solution 4.9.** (i) From the definition, the reversion of $a$ is $a^\dagger = \sigma_{654321}$.

Remark 4.2. Note that in this case $a^\dagger = -a$.

**Solution 4.10.** Its blades are $< a >_0 = 1$ and $< a >_2 = 2\sigma_{12} < a >_7 = \sigma_{123456}$.

**Solution 4.11.** $a \wedge b = -a$ then $I(a \wedge b) = -Ia$.

**Solution 4.12.** The norm of $a$ is.

$$\begin{aligned}
\|a\| &= \sqrt{\lfloor aa^\dagger \rfloor} \\
&= \sqrt{\lfloor (\sigma_1 + \sigma_2 + \cdots + \sigma_n)(\sigma_1 + \sigma_2 + \cdots + \sigma_n) \rfloor} \\
&= \sqrt{\lfloor n + \overline{\text{vector-residue}} \rfloor} \\
&= \sqrt{n}.
\end{aligned} \qquad (4.10)$$

**Solution 4.13.** (i) $a(bc) = \sigma_{12346}$. (ii) $(ab)c = \sigma_{12346}$. (iii) From the results (i) and (ii) yes, it is.

**Solution 4.14.** (i) $Ia = \sigma_{1234}a =$. (ii) $aI = a\sigma_{1234}$. (iii) From these results, (i) is a **rotation** of $\frac{\pi}{2}$ in the **clockwise** direction and (ii) is a **rotation** of $\frac{\pi}{2}$ in the **counter-clockwise** direction.

**Solution 4.15.** (i) $IIa = \sigma_{1234}\sigma_{1234}a$. (ii) $aII = a\sigma_{1234}\sigma_{1234}$. (iii) From these results, (i) is a **reflection** of $\pi$ in the **clockwise** direction and (ii) is a **reflection** of $\pi$ in the **counter-clockwise** direction.

**Solution 4.16.** $u = \dfrac{u}{||u||} = \dfrac{\sigma_1 - 2\sigma_2}{3}$. Then $y = -uxu = \dfrac{2}{3}\sigma_1 + \dfrac{8}{9}\sigma_2 + \sigma_3$.

**Solution 4.17.** The **line** $L_{x_0}(v)$ is given by

$$L_{x_0=(1,1,1,1)}(v) := \{\mathbf{x} \mid (\mathbf{x} - x_0) \wedge v = 0\}$$

$$[(x_1\sigma_1 + x_2\sigma_2 + x_3\sigma_3 + x_4\sigma_4) - (\sigma_1 + \sigma_2 + \sigma_3 + \sigma_4)] \wedge (\sigma_1 + \sigma_4) = 0$$
$$[(x_1 - 1)\sigma_1 + (x_2 + 1)\sigma_2 + (x_3 - 1)\sigma_3 + (x_4 - 1)\sigma_4] \wedge (\sigma_1 + \sigma_4) = 0$$

The exterior product $(x - x_0) \wedge v = \frac{1}{2}[(x - x_0)v - v(x - x_0)]$,

$$[(x_2 - 1)\sigma_{12} + (x_3 - 1)\sigma_{13} - (x_4 - 1)\sigma_{14}] = 0 \tag{4.11}$$

From (Eq. 4.11), $x_1 = \mathbb{R}, x_2 = 1, x_3 = 1$, and $x_4 = 0$. So, the points with the form $(\mathbb{R}, 1, 1, 1)$ are the solution. Note that the point $(1, 1, 1, 1)$ meets the **line** $L_{x_0}(v)$.

**Solution 4.18.** The **plane** $P_{x_0}(u, v)$ is given by

$$P_{x_0=(2,1,1,1,1)}(u \wedge v) := \{\mathbf{x} \mid (\mathbf{x} - x_0) \wedge (u \wedge v) = 0\} \tag{4.12}$$

$$[(x_1\sigma_1 + x_2\sigma_2) - (\sigma_1 + 2\sigma_2)] \wedge (\sigma_1\sigma_2) = 0$$
$$[(x_1 - 1)\sigma_1 + (x_2 - 2)\sigma_2] \wedge (\sigma_1\sigma_2) = 0 \tag{4.13}$$

From (Eq. 4.13), $[(x_1 - 1)\sigma_1 + (x_2 - 2)\sigma_2][\sigma_1\sigma_2] - [\sigma_1\sigma_2][(x_1 - 1)\sigma_1 + (x_2 - 2)\sigma_2] = 0$. So, the points $(x_2 - 2, x_1 - 1)$ are the solution. Note that the point $(2, 1, 1, 1, 1)$ meets the **plane** $P_{x_0=(2,1,1,1,1)}(u \wedge v)$.

## Solutions for Chapter 5

**Solution 5.1.** The $n$ degree of a $w_i$ form is the term that corresponds to the highest degree in the form. $w_o(x,y,z) = 3 + 2xyz$ is a $0-$form. $w_i(x,y,z) = 3 + 2xyz + 4dz$ is a $1-$form. $w_2(x,y,z) = 3 + 2xyz + 4dz + dydz$ is a $2-$form. Note this includes terms of a $0-$form and a $1-$form. $w_3(x,y,z) = 2 + e^{xyz}dx \wedge dy \wedge dz$. $w_4(x_1,x_2,x_3,x_4) = e^{x_1}dx_1 \wedge d_2 \wedge d_3 \wedge d_4$.

**Solution 5.2.**

$$
\begin{aligned}
d(e^{x^2yz}) &= (e^{x^2yz})'_x + (e^{x^2yz})'_y + (e^{x^2yz})'_z \\
&= 2xyze^{x^2yz} + x^2ze^{x^2yz} + x^2ye^{x^2yz}
\end{aligned}
\tag{5.14}
$$

**Solution 5.3.**

$$
\begin{aligned}
d(e^{x^2yz}dx + \sin xyzdy) &= d(e^{x^2yz}dx) + d(\sin xyzdy) \\
&= (e^{x^2yz})'_x dx + (e^{x^2yz})'_y dx + (e^{x^2yz})'_z dx \\
&\quad + (\sin xyz)'_x dx + (\sin xyz)'_y dx + (\sin xyz)'_z dx \\
&= x^2ze^{x^2yz}dydx + x^2ye^{x^2yz}dzdx \\
&\quad + xz\cos xyzdydx + xy\cos xyzdzdx
\end{aligned}
\tag{5.15}
$$

**Solution 5.4.**

$$
\begin{aligned}
d(x^2y + y^3) &= (x^2y + y^3)'_x + (x^2y + y^3)'_y + (x^2y + y^3)'_z \\
&= 2xy + 3y^2 + 0
\end{aligned}
\tag{5.16}
$$

**Solution 5.5.**

$$
\begin{aligned}
d(x^3y + y^3dydz) &= (x^3y + y^3dydz)'_x + (x^3y + y^3dydz)'_y + (x^3y + y^3dydz)'_z \\
&= 3x^2ydxdydz
\end{aligned}
\tag{5.17}
$$

**Solution 5.6.**

$$
\begin{aligned}
d\left(\frac{-x}{x^2y + y^2}dxdy\right) &= \left(\frac{-x}{x^2y + y^2}dxdy\right)'_x + \left(\frac{-x}{x^2y + y^2}dxdy\right)'_y + \left(\frac{-x}{x^2y + y^2}dxdy\right)'_z \\
&= 0\,dxdydz
\end{aligned}
\tag{5.18}
$$

**Solution 5.7.** (i)

$$dw = d(xdx + yzdy + x^3ydz)$$
$$= d(xdx) + d(yzdy) + d(x^3ydz)$$
$$= d(x)dx + d(yz)dy + d(x^3y)dz$$
$$= dxdx + zdydy + ydzdy + 3x^2ydxdz + x^3dydz$$
$$= ydzdy + 3x^2ydxdz + x^3dydz$$
$$= -ydydz + 3x^2ydxdz + x^3dydz \qquad (5.19)$$
$$= 3x^2ydxdz + (x^3 - y)dydz$$
$$= 3x^2ydx \wedge dz + (x^3 - y)dy \wedge dz$$

$$d(xdx) = (x)'_x dx + (x)'_y dx + (x)'_z dx$$
$$= dxdx + 0 + 0 \qquad (5.20)$$
$$= dxdx$$

$$d(yzdy) = (yz)'_x dy + (yz)'_y dy + (yz)'_z dy$$
$$= 0 + zdydy + ydzdy \qquad (5.21)$$

$$d(x^3ydy) = (x^3y)'_x dz + (x^3y)'_y dz + (x^3y)'_z dz$$
$$= 3x^2ydxdz + x^3dydz + 0 \qquad (5.22)$$

(ii)

$$d(dw) = d[3x^2ydxdz + (x^3 - y)dydz]$$
$$= d(3x^2ydxdz) + d[(x^3 - y)dydz]$$
$$= d(3x^2y)dxdz + d(x^3 - y)dydz \qquad (5.23)$$
$$= 6xydxdxdz + 3x^2dydxdz + 3x^2dxdydz - dydydz$$
$$= 0$$

$$d(3x^2y)dxdz = (3x^2y)'_x dxdz + (3x^2y)'_y dxdz + (3x^2y)'_z dxdz$$
$$= 6xydxdxdx + 3x^2dydxdz + 0 \qquad (5.24)$$

$$d(x^3 - y)dydz = (x^3 - y)'_x dydz + (x^3 - y)'_y dydz + (x^3 - y)'_z dydz$$
$$= 3x^2dxdydz - dydydz + 0 \qquad (5.25)$$

(iii)

$$(w \wedge \eta) = (xdx + yzdy + x^3ydz) \wedge (xydz)$$
$$= x^2ydxdz + y^2zxdydz + x^3yx^2dzdz \qquad (5.26)$$
$$= x^2ydxdz + y^2zxdydz$$

(iv)

$$
\begin{aligned}
d(w \wedge \eta) &= d(x^2 y dx dz + y^2 z x dy dz) \\
&= d(x^2 y dx dz) + d(y^2 z x dy dz) \\
&= d(x^2 y) dx dz + d(y^2 z x) dy dz \\
&= -x^2 dx dy dz + y^2 z dx dy dz \\
&= (y^2 z - x^2) dx dy dz
\end{aligned}
\tag{5.27}
$$

$$
\begin{aligned}
d(x^2 y) dx dz &= (x^2 y)'_x dx dz + (x^2 y)'_y dx dz + (x^2 y)'_z dx dz \\
&= 2xy dx dx dz + x^2 dy dx dz + 0 \\
&= -x^2 dx dy dz
\end{aligned}
\tag{5.28}
$$

$$
\begin{aligned}
d(y^2 z x) dy dz &= (y^2 z x)'_x dy dz + (y^2 z x)'_y dy dz + (y^2 z x)'_z dy dz \\
&= y^2 z dx dy dz + 2 y x z dy dy dz + y^2 x dz dy dz \\
&= y^2 z dx dy dz
\end{aligned}
\tag{5.29}
$$

(v)

From (iv) $d(w \wedge \eta) = dw \wedge \eta + (-1)^k w \wedge d\eta$.

$$
\begin{aligned}
d(e^{x^2 yz}) dx dy &= (e^{x^2 yz})'_x dx dy + (e^{x^2 yz})'_y dx dy + (e^{x^2 yz})'_z dx dy \\
&= -x^2 y e^{x^2 yz} dx dy dz
\end{aligned}
\tag{5.30}
$$

**Solution 5.8.** (i)

$$
\begin{aligned}
w_{11} \wedge w_{12} &= \frac{1}{2}(w_{11} w_{12} - w_{12} w_{11}) \\
&= (3dx + dy)(e^x dx + 2dy) \\
&= 3e^x dx \wedge dx + 6 dx \wedge dy + e^x dy \wedge dx + 2 dy \wedge dy \\
&= (6 - e^x) dx \wedge dy \\
&= (6 - e^x) dx dy
\end{aligned}
\tag{5.31}
$$

(ii)

$$
\begin{aligned}
d(6 - e^x) \wedge dx \wedge dy &= -e^x dx \wedge dx \wedge dy \\
&= 0
\end{aligned}
\tag{5.32}
$$

**Solution 5.9.**

$$
\begin{aligned}
dx \wedge dy &= (-r \sin\theta d\theta + \cos\theta dr) \wedge (r \cos\theta d\theta + \sin\theta dr) \\
&= -r^2 \sin\theta \cos\theta d\theta d\theta - r \sin^2\theta d\theta dr + r \cos^2\theta dr d\theta + \cos\theta \sin\theta dr dr \\
&= (-r \sin^2\theta - r \cos^2\theta) d\theta \wedge dr \\
&= r dr \wedge d\theta
\end{aligned}
\tag{5.33}
$$

**Solution 5.10.** (i)

$$w_4 \wedge w_4 = dx_1 dx_3 \tag{5.34}$$

(ii)

$$
\begin{aligned}
dw_4 &= d(dx_1 dx_3) \\
&= (dx_1 dx_3)'_{x_1} + (dx_1 dx_3)'_{x_2} + (dx_1 dx_3)'_{x_3} + (dx_1 dx_3)'_{x_4} \\
&= -dx_1 dx_2 dx_3 + dx_1 dx_3 dx_4
\end{aligned}
\tag{5.35}
$$

# Solutions for Chapter 6

**Solution 6.1.** $\int_D w_0 = \int_1^3 3x^2 + 2x = (17+6) - (1^3+2) = 23 - 3 = 20.$

**Solution 6.2.** $\int_D w_0 = \int_1^2 \int_1^2 \int_1^2 x^2 + 2xy - z = (2^2 + 8 - 2) - (1^3 + 4 - 1) = 10$
$-4 = 6.$

**Solution 6.3.** $\int_D w_1 = \int_0^2 x^4 dx + 3xy dy - z dz = \int_0^2 t^4(t)'_t + 3t^3(t^2)'_t - t(t^3)'_t dt =$
$\int_0^2 6t^4 + t^4 - 3t^3 dt = \left[\frac{7}{5}t^4 - \frac{3}{4}t^3\right] = \frac{13}{20}.$

**Solution 6.4.** $\oint_T F \circ T(t) \cdot T'(t) dt = \int_{-\pi}^{\pi} (-t^5, 2\sin t) \cdot (1, 4t^3) dt = \int_{-\pi}^{\pi} -t^5 +$
$8t^3 \sin t \, dt = 16\pi(\pi^2 - 6).$

**Solution 6.5.** $\int_D w_1 = \int_{-\pi}^{\pi} -yx dx + \cos x dy = \int_{-\pi}^{\pi} -t^5(t)'_t + \cos t(t^4)'_t dt = \int_{-\pi}^{\pi}$
$-t^5 + 4t^3 \cos t \, dt = 0.$

**Solution 6.6.** $\int_D w_1 = \int_{-\pi}^{\pi} x^4 + 2x \cos x \, dx = \frac{2}{5}\pi^5.$

**Solution 6.7.** $\int_D w_2 = \int_0^1 \int_0^{\frac{\pi}{2}} -y dx dy + x^2 dy dz \, d\theta dr = \int_0^1 \int_0^{\frac{\pi}{2}} -r \sin\theta \frac{\partial(x,y)}{\partial(r,\theta)}$
$+ r \cos\theta \frac{\partial(y,z)}{\partial(r,\theta)} d\theta dr = \int_0^1 \int_0^{\frac{\pi}{2}} -r^2 \sin\theta + r^2 \cos^2\theta \sin\theta \, d\theta dr = -\frac{2}{9}.$

Note 6.1. $\frac{\partial(x,y)}{\partial(r,\theta)} = r$, and $\frac{\partial(y,z)}{\partial(r,\theta)} = \sin\theta.$

**Solution 6.8.** Using $T(r,\theta) = (r\cos\theta, r\sin\theta, 3)$ with $\theta \in [0, 2\pi]$, $r \in [0, \sqrt{2}]$.

$$\oiint_S F \circ T(r,\theta) \cdot \eta(r,\theta) dS = \int_{\theta_1}^{\theta_2} \int_{r_1}^{r_2} F(T(r,\theta)) \cdot \frac{\partial T}{\partial r} \times \frac{\partial T}{\partial \theta} dr d\theta$$
$$= \int_0^{2\pi} \int_0^{\sqrt{2}} F(T(r,\theta)) \cdot \frac{\partial T}{\partial r} \times \frac{\partial T}{\partial \theta} dr d\theta \qquad (6.36)$$
$$= \int_0^{2\pi} \int_0^{\sqrt{2}} (r\cos\theta, r\sin\theta, 1) \cdot (0, 0, r) dr d\theta$$
$$= 2\pi.$$

**Solution 6.9.** $\iint_D w_2 = \int_0^1 \int_0^x x^3 + 2xy dy dx = \int_0^1 x^4 + x^3 dx = \frac{9}{20}.$

**Solution 6.10.**

$$\int_D w_3 = \int_0^2 \int_0^{2\pi} \int_0^{\pi} [xyz dz dy dx] dr d\theta d\phi =$$
$$\int_0^1 \int_0^{\pi} \int_0^{\frac{\pi}{2}} r\theta\phi \frac{\partial(z,y,x)}{\partial(r,\theta,\phi)} = \frac{\pi^4}{8}. \qquad (6.37)$$

Note 6.2. $\dfrac{\partial(z,y,x)}{\partial(r,\theta,\phi)} = 1$

**Solution 6.11.** $\displaystyle\iiint_D w_3 = \int_0^1 \int_0^1 \int_0^1 x^4 z + 2xy\, dx\, dy = \frac{3}{5}.$

**Solution 6.12.** $\displaystyle\iiiint_D w_4 = \int_0^\pi \int_0^{2\pi} \int_0^{3\pi} \int_0^{4\pi} x_1 x_2 x_3 x_4^2 dx_4\, dx_3\, dx_2\, dx_1 = 24\pi^9.$

## Solutions for Chapter 7

**Solution 7.1.** $\iiint_D w_0 = \int_0^1 \int_0^2 \int_0^3 xyz\,dz\,dy\,dx = \left(\frac{9}{2}\right)\left(\frac{4}{2}\right)\left(\frac{1}{2}\right) = \frac{9}{2}.$ In the Heaviside-Gibbs algebra, this integral represents a volume in the $\mathbb{R}^3$ space or an area in $\mathbb{R}^2$. In the Geometric algebra this is a $0-$form $w_0$ integral.

**Solution 7.2.** $\int_D dw_1 = \int_0^{2\pi} yx\,dx + 2zy\,dy + dz = \int_0^{2\pi} (\cos t \sin t (\cos t)'_t + 2\sin t$

$(\sin t)'_t)\,dt = \int_0^{2\pi}(-\sin^2 t\cos t + 2\sin t\cos t)dt.$ Now if $F(x,y) = (yx, 2zy, 1),\Rightarrow$

$F\circ C = (\cos t\sin t, 2\sin t, 1),$ then $\int_0^{2\pi}(\cos t\sin t, 2\sin t, 1)\cdot(-\sin t, \cos t, 0)\,dt.$ So both integrals are equivalent.

**Solution 7.3.** $\int_D w_2 = \int_0^1 \int_0^{2\pi} [2zdxdy + 3xdydz + 4ydzdx]\,d\theta dr = \int_0^1\int_0^{2\pi}\Big[4$

$\frac{\partial(x,y)}{\partial(r,\theta)} + r\cos x\frac{\partial(y,z)}{\partial(r,\theta)} + r\sin x\frac{\partial(z,x)}{\partial(r,\theta)}\Big]d\theta dr = 24\int_0^1\int_0^{2\pi} rd\theta dr = 24\pi.$

Note 7.3. $\frac{\partial(x,y)}{\partial(r,\theta)} = 6r,\ \frac{\partial(y,z)}{\partial(r,\theta)} = 0,$ and $\frac{\partial(z,x)}{\partial(r,\theta)} = 0.$

If $F(x,y,z) = (2x, 3y, 4z),\ T(r,\theta) = (r\cos\theta, r\sin t, 1), r\in[0,1], \theta\in[0,2\pi]$, its Jacobian is 24, then $24\int_0^1\int_0^{2\pi} rd\theta\,dr = 24\pi.$ So both integrals are equivalent.

**Solution 7.4.** $\int_{\partial D} w_1 = \int_0^{2\pi} -4y\,dx + 4x\,dy\ dt = \int_0^{2\pi} -4\sin t(4\cos t)'_t + 4\cos t$

$(4\sin t)'_t\,dt$

$= \int_0^{2\pi} 16\sin^2 t + 16\cos^2 t\,dt = 16\pi.$

$$dw_1 = d(-4y\wedge dx) + d(4x\wedge dy))$$

$$= -\left(\frac{\partial 4y}{\partial x}dx + \frac{\partial 4y}{\partial y}dy\right)\wedge dx + \left(\frac{\partial 4x}{\partial x}dx + \frac{\partial 4x}{\partial y}dy\right)\wedge dy$$

$$= -\left(\frac{\partial 4y}{\partial x}\right)dx\wedge dx - \left(\frac{\partial 4y}{\partial y}\right)dy\wedge dx \qquad (7.38)$$

$$+ \left(\frac{\partial 4x}{\partial x}\right)dx\wedge dy + \left(\frac{\partial 4x}{\partial y}\right)dy\wedge dy$$

$$= 8dxdy$$

Now, we parameterize $c(r,\theta) = (r\cos\theta, r\sin\theta, 1)$

$$\int_D dw_1 = \int_0^1\int_0^{2\pi}[8dxdy]\,d\theta dr = \int_0^1\int_0^{2\pi}\left[8\frac{\partial(x,y)}{\partial(r,\theta)}\right]d\theta dr = \int_0^1\int_0^{2\pi} 8rd\theta dr$$

$= 16\pi.$

Note 7.4. $\dfrac{\partial(x,y)}{\partial(r,\theta)} = r.$

Green's theorem is verified.

**Solution 7.5.**

$$\iint_D \left(\frac{\partial Q}{\partial x} - \frac{\partial P}{\partial y}\right) dy\, dx = \int_0^2 \int_0^{2-x} 2xy^2 - xdy\, dx$$
$$= -\frac{4}{15}$$

$$\int_{\partial D} P \circ c(t)\frac{x(t)}{dt} + Q \circ c(t)\frac{y(t)}{dt} = \int_{\partial D_1} P \cdot c_1(t)\frac{dx}{dt} + Q \cdot c_1(t)\frac{dy}{dt}$$
$$+ \int_{\partial D_2} P \cdot c_2(t)\frac{dx}{dt} + Q \cdot c_2(t)\frac{dy}{dt} \qquad (7.39)$$
$$+ \int_{\partial D_3} P \cdot c_3(t)\frac{dx}{dt} + Q \cdot c_3(t)\frac{dy}{dt}$$
$$= 0 - \frac{4}{15} + 0$$
$$= -\frac{4}{15}$$

Where $c_1(t) = (t,0), t \in [0,2]$, $c_2(t) = (2-t,t), t \in [0,2]$, and $c_3(t) = (0,t), t \in [2,0]$.

$$\int_{\partial D_1} P \cdot c_1(t)\frac{dx}{dt} + Q \cdot c_1(t)\frac{dy}{dt} = \int_0^2 xy(0) + x^2y^2(2)dt$$
$$= \int_0^2 0(1) + 0(0)dt$$
$$= 0$$

$$\int_{\partial D_2} P \cdot c_2(t)\frac{dx}{dt} + Q \cdot c_2(t)\frac{dy}{dt} = \int_0^2 xy(-1) + x^2y^2(1)dt$$
$$= \int_0^2 -t(1-t) + (1-t)^2(t^2)dt \qquad (7.40)$$
$$= -\frac{4}{15}$$

$$\int_{\partial D_3} P \cdot c_3(t)\frac{dx}{dt} + Q \cdot c_3(t)\frac{dy}{dt} = \int_2^0 xy(1) + x^2y^2(0)dt$$
$$= \int_2^0 0(0) + 0(1)dt$$
$$= 0$$

**Solution 7.6.** $\displaystyle\int_{\partial D} w_1 = \int_0^{2\pi} y\,dx + e^z\,dy + x\,dz\,dt = \int_0^{2\pi} \sin t (\cos t)'_t + e^1$

$\displaystyle(\sin t)'_t + \cos t(1)'_t\,dt = \int_0^{2\pi} -\sin^2 t + e\cos t\,dt = -\pi.$

$$dw_1 = d(y \wedge dx) + d(e^z \wedge dy) + d(x \wedge dz)$$

$$= \left(\frac{\partial y}{\partial x}dx + \frac{\partial y}{\partial y}dy + \frac{\partial y}{\partial z}dz\right) \wedge dx$$

$$+ \left(\frac{\partial e^z}{\partial x}dx + \frac{\partial e^z}{\partial y}dy + \frac{\partial e^z}{\partial z}dz\right) \wedge dy$$

$$+ \left(\frac{\partial x}{\partial x}dx + \frac{\partial x}{\partial y}dy + \frac{\partial x}{\partial z}dz\right) \wedge dz$$

$$\text{(7.41)}$$

$$= \left(\frac{\partial y}{\partial x}\right)dx \wedge dx + \left(\frac{\partial y}{\partial y}\right)dy \wedge dx + \left(\frac{\partial y}{\partial z}\right)dz \wedge dx$$

$$+ \left(\frac{\partial e^z}{\partial x}\right)dx \wedge dy + \left(\frac{\partial e^z}{\partial y}\right)dy \wedge dy + \left(\frac{\partial e^z}{\partial z}\right)dz \wedge dy$$

$$= \left(\frac{\partial x}{\partial x}\right)dx \wedge dz + \left(\frac{\partial x}{\partial y}\right)dy \wedge dz + \left(\frac{\partial x}{\partial z}\right)dz \wedge dz$$

$$= -dxdy + dzdx + e^z dzdy$$

Now, we parameterize $c(r,\theta) = (r\cos\theta, r\sin\theta, 1)$

$$\int_D dw_1 = \int_0^1 \int_0^{2\pi} [-dxdy + dzdx + e^z dzdy]\,d\theta dr = \int_0^1 \int_0^{2\pi} \left[-\frac{\partial(x,y)}{\partial(r,\theta)}\right.$$

$$\left.+\frac{\partial(z,x)}{\partial(r,\theta)} + e^1\frac{\partial(z,y)}{\partial(r,\theta)}\right]d\theta dr = \int_0^1 \int_0^{2\pi} -r\,d\theta dr = -\pi.$$

Note 7.5. $\dfrac{\partial(x,y)}{\partial(r,\theta)} = r,$ $\dfrac{\partial(z,x)}{\partial(r,\theta)} = 0,$ and $\dfrac{\partial(z,y)}{\partial(r,\theta)} = 0.$

Stokes's theorem is verified.

**Solution 7.7.** If $T(\theta,r) = (r\cos t, r\sin t, 1 - r\cos t - r\sin t),$

$$\frac{\partial(T_y, T_z)}{\partial(\theta,r)} = \begin{vmatrix} r\cos\theta & \sin\theta \\ r\sin\theta & -r\cos\theta \end{vmatrix} \tag{7.42}$$

$$\frac{\partial(T_z, T_x)}{\partial(\theta,r)} = \begin{vmatrix} r\sin\theta & -r\cos\theta \\ -r\sin\theta & \sin\theta \end{vmatrix} \tag{7.43}$$

$$\frac{\partial(T_x, T_y)}{\partial(\theta,r)} = \begin{vmatrix} -r\sin\theta & \sin\theta \\ r\cos\theta & \sin\theta \end{vmatrix} \tag{7.44}$$

$$\iint_S dw = \int_0^1 \int_0^{2\pi} (\frac{\partial R}{\partial y} - \frac{\partial Q}{\partial z}) dydz + (\frac{\partial P}{\partial z} - \frac{\partial R}{\partial x}) dzdx + (\frac{\partial Q}{\partial x} - \frac{\partial P}{\partial y}) dxdy$$

$$= \int_0^1 \int_0^{2\pi} (\frac{\partial R}{\partial y} - \frac{\partial Q}{\partial z}) \circ T(\theta, r) \frac{\partial(T_y, T_z)}{\partial(\theta, r)}$$

$$+ (\frac{\partial P}{\partial z} - \frac{\partial R}{\partial x}) \circ T(\theta, r) \frac{\partial(T_z, T_x)}{\partial(\theta, r)}$$

$$+ (\frac{\partial Q}{\partial x} - \frac{\partial P}{\partial y}) \circ T(\theta, r) \frac{\partial(T_x, T_y)}{\partial(\theta, r)}$$

$$= \int_0^1 \int_0^{2\pi} (0) \frac{\partial(T_y, T_z)}{\partial(\theta, r)} + (0) \frac{\partial(T_z, T_x)}{\partial(\theta, r)} + (0) \frac{\partial(T_x, T_y)}{\partial(\theta, r)} d\theta dr$$

$$= 0$$

$$(7.45)$$

$$\int_{\partial D} P \cdot c(t) \frac{dx}{dt} + Q \cdot c(t) \frac{dy}{dt} + R \cdot c(t) \frac{dy}{dt} = \int_0^{2\pi} (\cos t)(-\sin t)$$

$$+ (\sin t)(\cos t)$$

$$+ (1 - \cos t - \sin t)(\sin t - \cos t) dt$$

$$= 0$$

$$(7.46)$$

**Solution 7.8.** $\int_D w_2 = \int_0^\pi \int_0^{2\pi} [xz\,dxdy - xy\,dxdz - dydz]\,d\theta d\phi = \int_0^\pi \int_0^{2\pi} \Big[$

$\cos\theta \sin\phi \cos\phi \frac{\partial(y,x)}{\partial(r,\theta)} + \cos\theta \sin\phi \sin\theta \sin\phi \frac{\partial(z,x)}{\partial(r,\theta)} + \frac{\partial(z,y)}{\partial(r,\theta)}\Big] d\theta d\phi$

$= \int_0^\pi \int_0^{2\pi} d\theta d\phi = -\cos\theta \sin\phi \cos\phi \sin\phi \cos\phi - \cos\theta \sin\phi \sin\theta \sin\phi \sin^2\phi \sin\theta$

$- \sin^2\phi \cos\theta\, d\theta\, d\phi = -\int_0^\pi \int_0^{2\pi} \cos\theta \cos^2\phi \sin^2\phi + \cos\theta \sin^3\phi \sin^2\theta \cos\theta$

$+ \sin^2\phi\, d\theta\, d\phi = 0.$

**Note 7.6.** $\frac{\partial(x,y)}{\partial(\theta,\phi)} = -\sin\phi\cos\phi, \frac{\partial(z,x)}{\partial(\theta,\phi)} = -\sin^2\phi\sin\theta,$ and $\frac{\partial(z,y)}{\partial(\theta,\phi)} = \sin^2\phi$
$\cos\theta.$

$$dw_2 = -d(dydz) + d(xydzdx) + d(xzdxdy)$$

$$
\begin{aligned}
= &(-\frac{\partial 1}{\partial x}dx - \frac{\partial 1}{\partial y}dy - \frac{\partial 1}{\partial z}dz) \wedge (dy \wedge dz) \\
&+ (\frac{\partial xy}{\partial x}dx + \frac{\partial xy}{\partial y}dy + \frac{\partial xy}{\partial z}dz) \wedge (dz \wedge dx) \\
&+ (\frac{\partial xz}{\partial x}dx + \frac{\partial xz}{\partial y}dy + \frac{\partial xz}{\partial z}dz) \wedge (dx \wedge dy)
\end{aligned}
$$

$$
\begin{aligned}
= &-\frac{\partial 1}{\partial x}dx(dy \wedge dz) - \frac{\partial 1}{\partial y}dy(dy \wedge dz) - \frac{\partial 1}{\partial z}dz(dy \wedge dz) \\
&+ \frac{\partial xy}{\partial x}dx(dz \wedge dx) + \frac{\partial xy}{\partial y}dy(dz \wedge dx) + \frac{\partial xy}{\partial z}dz(dz \wedge dx) \\
&+ \frac{\partial xz}{\partial x}dx(dx \wedge dy) + \frac{\partial xz}{\partial y}dy(dx \wedge dy) + \frac{\partial xz}{\partial z}dz(dx \wedge dy)
\end{aligned}
\tag{7.47}
$$

$$
\begin{aligned}
= &-\frac{\partial 1}{\partial x}dxdydz - \frac{\partial 1}{\partial y}dydydz - \frac{\partial 1}{\partial z}dzdydz \\
&+ \frac{\partial xy}{\partial x}dxdzdx + \frac{\partial xy}{\partial y}dydzdx + \frac{\partial xy}{\partial z}dzdzdx \\
&+ \frac{\partial xz}{\partial x}dxdxdy + \frac{\partial xz}{\partial y}dydxdy + \frac{\partial xz}{\partial z}dzdxdy
\end{aligned}
$$

$$= 2x\,dxdydz$$

Ahora aplicamos la parametrizacion $T(\rho, \theta, \phi) = (\rho \cos \theta \sin \phi, \rho \sin \theta \sin \phi, \rho \cos \phi)$, $\theta \in [0, 2\pi]$, $\phi \in [0, \pi]$, and $\rho \in [0, 1]$.

$$
\int_D dw_2 = \int_0^1 \int_0^\pi \int_0^{2\pi} 2x dx dy dz\, d\theta\, d\phi\, d\rho = \int_0^1 \int_0^\pi \int_0^{2\pi} 2\rho \cos \theta \sin \phi
$$
$$
\frac{\partial(x,y,z)}{\partial(\rho,\theta,\phi)} = \int_0^1 \int_0^\pi \int_0^{2\pi} \rho \cos\phi 2\rho \cos \theta \sin\phi d\theta\, d\phi\, d\rho = 2\rho^2 \cos \phi \sin \phi
$$
$$
d\phi\, d\rho = 0.
$$

Note 7.7. $\dfrac{\partial(x,y,z)}{\partial(\rho,\theta,\phi)} = \rho \cos \phi.$

Gauss's theorem is verified.

**Solution 7.9.**

$$
\begin{aligned}
\int_\Omega &= \iiint_\Omega (\frac{\partial P}{\partial x} + \frac{\partial Q}{\partial y} + \frac{\partial R}{\partial z}) dz dy dx \\
&= \int_{-1}^{-1} \int_{-\sqrt{1-x^2}}^{\sqrt{1-x^2}} \int_{-\sqrt{1-x^2-y^2}}^{\sqrt{1-x^2-y^2}} 8xyz\, dz dy dx \\
&= 0
\end{aligned}
\tag{7.48}
$$

$$\iint_{\partial\Omega} d\Omega = \int_0^\pi \int_0^{2\pi} P \circ T \frac{\partial(T_y, T_z)}{\partial(\theta, \phi)} + Q \circ T \frac{\partial(T_z, T_x)}{\partial(\theta, \phi)} + R \circ T \frac{\partial(T_x, T_y)}{\partial(\theta, \phi)} d\theta \, d\phi$$

$$= \int_0^\pi \int_0^{2\pi} \cos^2\theta \sin^2\phi \frac{\partial(T_y, T_z)}{\partial(\theta, \phi)} + \sin^2\theta \sin^2\phi \frac{\partial(T_z, T_x)}{\partial(\theta, \phi)} + \cos^2\phi$$

$$\frac{\partial(T_x, T_y)}{\partial(\theta, \phi)} d\theta d\phi$$

$$= \int_0^\pi \int_0^{2\pi} (\cos^2\theta \sin^2\phi)(-\cos\theta \sin^2\phi) + (\sin^2\theta \sin^2\phi)(-\sin^2\phi)$$

$$+ (\cos^2\phi)(-\sin^2\theta \sin\phi \cos\phi - \cos^2\theta \sin\phi \cos\phi) d\theta d\phi$$

$$= 0$$

$$(7.49)$$

Note 7.8. The sign depends on the orientation.

$$\frac{\partial(T_y, T_z)}{\partial(\theta, r)} = \begin{vmatrix} \cos\theta\cos\phi & \sin\theta\cos\phi \\ 0 & -\sin\phi \end{vmatrix}$$
$$= -\cos\theta\sin^2\phi$$

$$(7.50)$$

$$\frac{\partial(T_z, T_x)}{\partial(\theta, r)} = \begin{vmatrix} 0 & -\sin\phi \\ -\sin\theta\sin\phi & \cos\theta\cos\phi \end{vmatrix}$$
$$= -\sin^2\phi$$

$$(7.51)$$

$$\frac{\partial(T_x, T_y)}{\partial(\theta, r)} = \begin{vmatrix} -\sin\theta\sin\phi & \cos\theta\cos\phi \\ \cos\theta\sin\phi & \sin\theta\cos\phi \end{vmatrix}$$
$$= -\sin^2\theta\sin\phi\cos\phi - \cos^2\theta\sin\phi\cos\phi$$

$$(7.52)$$

**Solution 7.10.**

$$\int_D w_3 = \int_0^1 \int_0^2 \int_0^3 [x_3 x_4 \, dx_1 \, dx_2 \, dx_3] \, du_1 du_2 du_3$$

$$= \int_0^1 \int_0^2 \int_0^3 \left[ u_1 u_3 \frac{\partial(x_1, x_2, x_3)}{\partial(u_1, u_2, u_3)} \right] du_1 du_2 du_3$$

$$= \int_0^1 \int_0^2 \int_0^3 u_1 u_3 du_1 du_2 du_3$$

$$= \frac{9}{2}$$

$$(7.53)$$

Note 7.9. $\dfrac{\partial(x_1, x_2, x_3)}{\partial(u_1, u_2, u_3)} = 1$.

$$dw_3 = d(x_1 x_4 \, dx_1 dx_2 dx_3 + x_2 x_3 \, dx_3 dx_4 dx_1)$$

$$= (\frac{\partial x_1 x_4}{\partial x_1} dx_1 + \frac{\partial x_1 x_4}{\partial x_2} dx_2 + \frac{\partial x_1 x_4}{\partial x_3} dx_3 + \frac{\partial x_1 x_4}{\partial x_4} dx_4) \wedge (dx_1 \wedge dx_2 \wedge dx_3)$$
$$+ (\frac{\partial x_2 x_3}{\partial x_1} dx_1 + \frac{\partial x_2 x_3}{\partial x_2} dx_2 + \frac{\partial x_2 x_3}{\partial x_3} dx_3 + \frac{\partial x_2 x_3}{\partial x_4} dx_4) \wedge (dx_3 \wedge dx_4 \wedge dx_1)$$

$$= \frac{\partial x_1 x_4}{\partial x_1} dx_1(dx_1 \wedge dx_2 \wedge dx_3) + \frac{\partial x_1 x_4}{\partial x_2} dx_2(dx_1 \wedge dx_2 \wedge dx_3)$$
$$+ \frac{\partial x_1 x_4}{\partial x_3} dx_3(dx_1 \wedge dx_2 \wedge dx_3) + \frac{\partial x_1 x_4}{\partial x_4} dx_4(dx_1 \wedge dx_2 \wedge dx_3)$$
$$= \frac{\partial x_2 x_3}{\partial x_1} dx_1(dx_3 \wedge dx_4 \wedge dx_1) + \frac{\partial x_2 x_3}{\partial x_2} dx_2(dx_3 \wedge dx_4 \wedge dx_1)$$
$$+ \frac{\partial x_2 x_3}{\partial x_3} dx_3(dx_3 \wedge dx_4 \wedge dx_1) + \frac{\partial x_2 x_3}{\partial x_4} dx_4(dx_3 \wedge dx_4 \wedge dx_1)$$

$$= \frac{\partial x_1 x_4}{\partial x_1} dx_1 dx_1 dx_2 dx_3 + \frac{\partial x_1 x_4}{\partial x_2} dx_2 dx_1 dx_2 dx_3$$
$$+ \frac{\partial x_1 x_4}{\partial x_3} dx_3 dx_1 dx_2 dx_3 + \frac{\partial x_1 x_4}{\partial x_4} dx_4 dx_1 dx_2 dx_3$$
$$+ \frac{\partial x_2 x_3}{\partial x_1} dx_1 dx_3 dx_4 dx_1 + \frac{\partial x_2 x_3}{\partial x_2} dx_2 dx_3 dx_4 dx_1$$
$$+ \frac{\partial x_2 x_3}{\partial x_3} dx_3 dx_3 dx_4 dx_1 + \frac{\partial x_2 x_3}{\partial x_4} dx_4 dx_3 dx_4 dx_1$$

$$= -(x_1 + x_3) \, dx_1 dx_2 dx_3 dx_4$$

$$(7.54)$$

Now, we parameterize $T(u_1, u_2, u_3, u_4) = (u_1, u_2, u_3, u_4), u_1 \in [0, \alpha_1]$, $u_2 \in [0, \alpha_2]$, $u_3 \in [0, \alpha_3]$, and $u_4 \in [0, \alpha_4]$.

$$\int_D dw_3 = \int_0^{\alpha_1} \int_0^{\alpha_2} \int_0^{\alpha_3} \int_0^{\alpha_4} (x_1 + x_3) \, dx_1 dx_2 dx_3 dx_4 = \frac{9}{2}.$$ Where $\alpha_1 = 1, \alpha_2 = 2, \alpha_3 = 3$, and $\alpha_4 = -1 - \frac{-\sqrt{7}}{2}$, or $\alpha_4 = \frac{1}{2}\left(\sqrt{7} - 2\right)$.

The Fundamental Theorem of Calculus is verified.

# References

[1] C. Polanco, *Advanced Calculus: Fundamentals of Mathematics.*    Bentham Science Publishers, -Sharjah, UAE. ISBN 9789811415081, 2019.

[2] E. Chisolm, "Geometric algebra," *arXiv*, vol. 5935, no. 205, pp. 1–92, 2012.

[3] E. Catsigerasi, "Derivada exterior de formas diferenciales." 2020. [Online]. Available: https://www.youtube.com/watch?v=cwoLDV8NkFI

[4] J. Marsden and A. Tromba, *Vector Calculus.*    New York, NY 10004, USA: W H Freeman And Company, 2011.

[5] J. Hefferson, *Linear Algebra.*    Colchester, VT 05439, USA: Saint Michael's College, 2014, http://joshua.smcvt.edu/linearalgebra/BOOK.pdf.

[6] A. Malcev, *Foundations of Linear Algebra.*    New York, NY 10004, USA: W H Freeman And Company, 1963.

[7] W. Rudin, *Principles of Mathematical Analysis.* New York, NY 10020, USA: McGraw-Hill, 1964, https://notendur.hi.is/vae11/%C3%9Eekking/principles_of_mathematical_a nalysis_walter_rudin.pdf.

[8] W. S. Massey, "Cross products of vectors in higher dimensional euclidean spaces," *The American Mathematical Monthly*, vol. 90, no. 10, pp. 697–701, 12 1983.

[9] R. Ablamowicz and G. Sobczyk, "Lecture series of clifford algebras and their applications," May 18 2002.

[10] C. Pereyda-Pierre and A. Castellanos-Moreno, "La derivada geométrica y el cálculo geométrico." *Memorias de la Semana de Investigación y Docencia en Matemáticas, Universidad Autónoma de Sonora*, vol. 60, no. 4, pp. 115–120, 5.

[11] J. Rotman, *Advanced Modern Algebra.*    Upper Saddle River, NJ 07458, USA: Pearson Education, 2002.

[12] D. Cherney, T. Denton, R. Thomas, and A. Waldron, *Linear Algebra*, Katrina Glaeser and Travis Scrimshaw., Ed.    Davis, CA 95616, USA: Creative Commons Attribution-NonCommercial-ShareAlike 3.0 Unported License, 2013, https://www.math.ucdavis.edu/ linear/linear-guest.pdf.

[13] R. Beezer, *A first course of Linear Algebra.*    Tacoma, WA 98416, USA: University of Puget Sound, 2017, http://linear.ups.edu/jsmath/0220/fcla-jsmath-2.20li61.html.

[14] F. A. Jr. and E. Mendelson, *CALCULUS.*    SCHAUM'S outlines McGraw-Hill, -USA.

[15] R. Bartle and D. Sherbert, *Introduction to Real Analysis.*    John & Wiley Sons, Inc., USA.

[16] C. J. L. Doran, "Geometric algebra and its application to mathematical physics," Ph.D. dissertation, Sidney Sussex College, University of Cambridge, 1994.

[17] A. Castellanos-Moreno, *Introducción al Álgebra y al Cálculo Geométrico.* Departamento de Física, Universidad Autónoma de Sonora, 2013.

[18] M. R. Spiegel, *Theory and Problems of Advanced Calculus.*    Schaum Publishing CO. New York, N.Y. U.S.A., 1967.

[19] G. Wilkin, "Examples of stokes' theorem and gauss' divergence theorem," 2018, http://www.math.jhu.edu/g̃raeme/files/math202_spring2009/StokesandGau ss.pdf.

[20] J. Smith, "Making sense of adding bivectors," https://mx.linkedin.com/in/james-smith-1b195047, 2018.

[21] M. Brittenham, "The surface area of a torus (i.e, doughnut)." 2012, https://www.math.unl.edu/m̃brittenham2/classwk/208s12/inclass/surface.ar ea.of.a.torus.pdf.

[22] S. Ramos, J. A. Juárez, and G. Sobczyk, "From vectors to geometric algebra 4, page 17." 2018, https://arxiv.org/pdf/1802.08153.pdf.

[23] J. Suter, "Geometric algebra primer," http://www.jaapsuter.com/geometric-algebra.pdf, 2003.

[24] S. Ramos-Ramírez and J. A. Juárez-González, "De vectores al álgebra geométrica," https://www.garretstar.com/GA3mex%20(ESP).pdf, 2018.

[25] J. Vince, *Geometric Algebra: An Algebraic System for Computer Games and Animation.* Springer Dordrecht Heidelberg London New York.

[26] Wikipedia, "Intersection of a line and a plane," 2018, https://en.wikipedia.org/wiki/Geometric_algebra.

[27] "Exterior derivative," 2020. [Online]. Available: https://en.wikipedia.org/wiki/Exterior˙derivative

[28] S. Schmit and S. Grützmacher, "Differential forms in $\mathbb{R}^n$," 2015, https://www.mathi.uni-heidelberg.de/lee/Stehpan_Sven.pdf.

[29] W. G. Faris, "Vector fields and differential forms," 2008, http://math.arizona.edu/ faris/mathanalweb/manifold.pdf.

[30] G. Ippolito, S. Lanini, P. Brouqui, A. Di caro, and F. Vairo, "Ebola: missed opportunities for europe–africa research," *Lancet Infect Dis*, vol. 15, pp. 1254–1255, 2015.

[31] J. Castañón González, C. Polanco, R. González González, and J. Carrillo ruiz, "Surveillance system for acute severe infections with epidemic potential based on a deterministic-stochastic model, the stochcum method," *Cirugía y Cirujanos*, vol. In press, 2020.

[32] R. Watkins, S. Eagleson, B. Veenendaal, G. Wright, and A. Plant, "Applying cusum-based methods for the detection of outbreaks of ross river virus disease in western australia," *BMC Med Inform Decis Mak*, 2008.

[33] Wikipedia, "Ley de ampere — wikipedia, la enciclopedia libre," 2020, [Internet; descargado 29-agosto-2020]. [Online]. Available: https://es.wikipedia.org/w/index.php?title=Ley˙de˙Amp%C3%A8re&oldid =128381461

# Subject Index

www.ingramcontent.com/pod-product-compliance
Lightning Source LLC
Chambersburg PA
CBHW041715210326
41598CB00007B/658